COPYRIGHT

ISBN Paperback 978-1-970503-23-4

ABOUT THE AUTHOR

Alexandria Isaacs

Isaacs is a scientific researcher with decades of experience working in facilities serving older adults, those with chronic conditions, and caregiving families. Her background gives her deep insight into care, aging, and quality-of-life issues. Now focused on research and writing, she explores technology, human wellbeing, and ethical AI, combining her clinical expertise with interests in robotics and future care. Living in Rural America, Isaacs brings an informed perspective on how tech and automation impact non-urban communities. Her blend of hands-on experience and innovative analysis makes her work credible and unique, bridging technical advancements with genuine human needs.

ABSTRACT

This book examines the ongoing transformation brought about by humanoid robots—a subtle yet significant revolution that is poised to reshape homes, workplaces, and communities in both practical and meaningful ways. It provides a framework for understanding the nature of humanoid robots, their anticipated arrival, and the impact they will have on various aspects of human life, including aging, employment, parenting, and interpersonal connections.

The perspective offered herein is not backed solely by technical data, but also by experience with real human needs, ethical considerations, and a commitment to exploring the responsible integration of technology into society. The content addresses essential questions regarding the capabilities of humanoid robots, implications for labor markets and economic stability, their potential role in caregiving while maintaining human dignity, the effects of robot companions on child development, and the relevance to rural communities, which is frequently excluded from futuristic narratives. Above all, this book considers how technology can be deployed to augment rather than diminish our humanity.

Rather than promoting exaggerated expectations or fear, this book serves as a guide for navigating a future in which humanoid robots become part of everyday life—as neighbors, colleagues, and caregivers. It is intended for parents preparing their children for a technologically advanced future, individuals caring for older adults, professionals concerned with workplace evolution, and readers seeking informed perspectives on imminent changes to their own lives.

The humanoid revolution is underway and is emerging not just in modern urban centers but surprisingly in environments with the greatest need—the homes of isolated elders, understaffed healthcare institutions, and communities contending with workforce shortages. Readers are invited to explore this evolving landscape with careful consideration, openness, and the collective wisdom necessary to guide its development responsibly.

TABLE OF CONTENTS

INTRODUCTION i

CHAPTER 1. WHAT EXACTLY IS A HUMANOID ROBOT? DEFINING THE NEW COMPANIONS 2

Distinguishing Humanoid Robots from AI, Automation, and Other Robotics: What Makes a Robot "Humanoid"? 3

Why Human Form Matters: The Psychology and Practicality of Building Robots That Look Like Us 7

The Current Reality: What Today's Humanoid Robots Can (and Cannot) Actually Do 10

CHAPTER 2. THE RISE OF HUMANOID ROBOTS: FROM SCIENCE FICTION TO YOUR FRONT DOOR 17

From Rosie to Reality: How Science Fiction Shaped Our Expectations (and Why Progress Took Longer Than We Thought) 18

The Breakthrough Decade: What Changed to Make Humanoid Robots Suddenly Possible 21

Who's Building Them and Why Now: The Companies, Research Labs, and Real-World Needs Driving the Humanoid Revolution 23

CHAPTER 3. HOW ROBOTS WILL CHANGE THE WORLD OF WORK: JOBS, DISPLACEMENT, AND ADAPTATION 29

The Honest Assessment: Which Jobs Are Most Vulnerable to Humanoid Automation (and Which Aren't) 30

Beyond Replacement: How Human-Robot Collaboration Will Create Hybrid Roles and Augment Human Capabilities 33

Preparing for the Shift: Practical Strategies for Workers, Families, and Communities to Adapt and Thrive 36

CHAPTER 4. HUMANOID ROBOTS IN EVERYDAY LIFE: TRANSFORMING HOMES, ROUTINES, AND DAILY LIVING 43

From Chores to Companionship: What Humanoid Robots Can (and Can't) Actually Do in Your Home 44

Aging in Place with Robotic Support: How Humanoids Could Help Seniors Maintain Independence and Dignity.. 47

The Privacy Paradox: Cameras, Data, and Trust When Robots Live Where You Live 49

CHAPTER 5. ROBOTS AND ELDER CARE IN THE FUTURE: DIGNITY, INDEPENDENCE, AND COMPASSIONATE TECHNOLOGY ... 56

The Elder Care Crisis: Why We Desperately Need New Solutions (and Why Robots Aren't the Only Answer)... 58

What Humanoid Robots Can Actually Do for Aging Adults: From Safety Monitoring to Cognitive Engagement and Companionship.................................. 60

The Ethics of Care: Ensuring Robot Assistance Preserves Dignity, Choice, and Human Connection Rather Than Replacing It 62

CHAPTER 6. CHILDREN AND HUMANOID ROBOTS: GROWING UP IN A ROBOT-INTEGRATED WORLD 69

The Developmental Question: How Robot Interaction Affects Social Skills, Empathy, and Human Relationship Formation in Children 70

Humanoid Robots as Educational Tools: Personalized Learning, Special Needs Support, and the Risk of Deepening Inequality 73

Parenting in the Robot Age: Setting Boundaries, Teaching Critical Thinking, and Ensuring Technology Serves Childhood Rather Than Replacing It.............. 76

CHAPTER 7. THE ROBOT REVOLUTION IN RURAL AMERICA: SMALL TOWNS, FARMS, AND UNDERSERVED COMMUNITIES............. 82

Agriculture's Robot Future: How Humanoid Robots Could Transform Farm Labor, Address Dangerous Conditions, and Help Aging Farmers Sustain Their Land 84

Healthcare Deserts and Service Gaps: Using Humanoid Robots to Bring Medical Support, Elder Care, and Essential Services to Underserved Rural Communities 87

The Infrastructure Challenge: Overcoming Barriers of Cost, Connectivity, and the Digital Divide to Ensure Rural America Isn't Left Behind 91

CHAPTER 8. ETHICAL IMPLICATIONS OF HUMANOID ROBOTS: RIGHTS, RESPONSIBILITIES, AND MORAL BOUNDARIES...........................99

The Rights Question: Do Humanoid Robots Deserve Protections, and What Happens When Machines Seem to Suffer?...............................100

Accountability and Responsibility: Who's to Blame When Robots Cause Harm—Manufacturers, Owners, or the Machines Themselves?...........................103

Moral Boundaries in Human-Robot Relationships: Consent, Manipulation, Dignity, and the Ethics of Emotional Attachment to Machines 106

CHAPTER 9. SAFETY, REGULATION, AND PREPARING FOR A ROBOT FUTURE: WHAT WE MUST DO NOW...........................112

The Safety Imperative: Physical Risks, Cybersecurity Vulnerabilities, and Why We Need Robot Safety Standards Now (Not Later) 114

Regulation Without Stifling Innovation: What Frameworks, Policies, and Oversight Structures Should Govern Humanoid Robots...........................117

Personal and Community Preparation: Practical Steps Individuals, Families, Schools, and Towns Can Take Today to Ready Themselves for Tomorrow's Robot Integration...........................119

CHAPTER 10. HOW AI AND ROBOTICS WILL TRANSFORM DAILY LIVING: MAINTAINING HUMANITY IN A HUMANOID WORLD...........................126

The Humanity Paradox: What We Risk Losing (and What We Might Gain) When Robots Handle Life's Daily Tasks 127

Intentional Living in a Robot World: Setting Boundaries, Protecting Connection, and Designing Technology Integration That Serves Human Flourishing...........................130

The Irreplaceable Human: Cultivating the Capacities, Relationships, and Experiences That Will Matter Most in an Automated Age...........................132

CONCLUSION...........................137

YOUR VOICE TRULY MATTERS...........................139

BIBLIOGRAPHY...........................140

INTRODUCTION

The world stands at the threshold of a transformation as profound as the industrial revolution or the dawn of the internet age. Humanoid robots—machines built in the human form, capable of walking, grasping, and interacting in ways that mirror human behavior—are moving from research laboratories into warehouses, hospitals, and before too long…living rooms, as well. Unlike previous technological shifts that changed how people work or communicate, this revolution will alter the most intimate dimensions of daily existence: who cares for aging parents, who teaches children, and who stocks grocery shelves in small towns.

This shift is arriving not as distant science fiction but as immediate reality. Companies across the globe are deploying humanoid robots in manufacturing facilities and logistics centers right now. Prototypes assist elderly individuals with mobility challenges and medication reminders. Researchers test humanoid tutors in classrooms and as companions in nursing homes. The question is no longer whether humanoid robots will become part of everyday life, but when—and whether society will be prepared for the practical, economic, and ethical challenges they bring.

The confusion surrounding humanoid robots runs deep. Many people conflate them with artificial intelligence software, industrial robot arms, or autonomous vehicles and completely miss what makes humanoid robots distinctive and consequential. A humanoid robot is not simply AI with a body, nor is it just another form of automation. A humanoid robot represents the convergence of artificial intelligence, advanced robotics, and human-centered design in a package specifically shaped to navigate human environments, use human tools, and interact with people in intuitively recognizable ways. This human form is not cosmetic; it will integrate into homes designed for human bodies, workplaces built around human capabilities, and social structures rooted in human-to-human interaction.

The promise these machines offer is substantial. In an aging world facing caregiver shortages, humanoid robots could help elderly individuals maintain independence and dignity in their own homes. In rural communities struggling with healthcare deserts and labor shortages, they could deliver essential services that geography and economics currently make impossible. For workers in dangerous industries, they could assume risks that previously would have cost human lives. For families overwhelmed by caregiving responsibilities, there is the opportunity to provide respite and support that preserves both the caregiver's wellbeing and the care recipient's quality of life.

Yet promise alone does not constitute preparation. The arrival of humanoid robots raises urgent questions that individuals, families, and communities must grapple with now, not after these machines have already reshaped society. Which jobs will disappear? Which will transform? And how can workers adapt before being displaced? How should parents navigate childhood development when robot companions and tutors become commonplace? What safety standards and regulatory frameworks must be established before humanoid robots enter vulnerable spaces like nursing homes and schools? When does robot assistance in elder care cross the line from supporting dignity to replacing irreplaceable human connection?

The ethical territory is equally complex. As humanoid robots become more sophisticated, questions emerge about who would be accountable when they cause harm, what the psychological impact could be of forming emotional attachments to machines, and whether society risks losing essential human traits—empathy, creativity, spontaneous connection—by outsourcing too much of life to automated helpers. These are not abstract philosophical puzzles but practical dilemmas that will shape family dynamics, community cohesion, and individual identity.

This book serves as a roadmap for navigating these questions with clarity, compassion, and practical wisdom. It distinguishes hype from reality, explores both opportunities and risks with unflinching honesty, and focuses on the human experience—particularly the perspectives of caregivers, aging adults, children, rural communities, and workers facing displacement. The goal is not to celebrate or condemn humanoid robots, but to prepare readers to live alongside them thoughtfully, to advocate for their responsible development, and to ensure that as these machines walk among us, they serve us humans without diminishing what actually makes us human.

CHAPTER 1. WHAT EXACTLY IS A HUMANOID ROBOT? DEFINING THE NEW COMPANIONS

The first time I see a humanoid robot hand a cup of water to an elderly patient in a rehabilitation facility, will surely shift my understanding of what "care" means. This isn't a science fiction fantasy or a distant future promise. This scenario is going to happen in the clinics very much like the ones where I've spent years as a biomedical engineer. I hope the patient's face will not show fear, but relief. That moment—simple, quiet, profoundly human in its need yet delivered by mechanical hands—will capture both the promise and the complexity of the humanoid revolution unfolding around us right now.

Most people have heard the term "humanoid robot" thrown around in the news, tech announcements, and science fiction stories, but few can articulate what actually makes a robot "humanoid" or how these machines differ from the AI assistants already living in our phones and smart appliances. The confusion is understandable. We're living through a period of breathtaking technological convergence where artificial intelligence, robotics, machine learning, and automation are advancing simultaneously, their boundaries blurring in ways that leave even experts struggling to explain where one technology ends and another begins. When a voice assistant answers questions, a self-driving car navigates traffic, a surgical robot performs delicate procedures, or a humanoid robot walks across a room, are these all the same thing? Are they all "AI"? Are they all "robots"? And if not, what distinguishes them from each other?

This chapter exists to clear away that confusion. Before we can thoughtfully explore how humanoid robots will transform work, caregiving, childhood, rural communities, and daily life and before we can grapple with the ethical questions and safety concerns they raise, we need clear definitions. We need to understand not just what humanoid robots are technically, but why humans have been driven for decades to create machines in our own image, what psychological and practical forces make the human form so compelling for robotics, and what today's humanoid robots can actually do versus what remains firmly in the realm of future possibility or outright fantasy.

The answers matter because humanoid robots aren't arriving in some distant, abstract future. They're being tested in warehouses and hospitals right now. They're being developed by major technology companies and well-funded startups with explicit plans to bring them into homes, workplaces, and communities within the next several years. Some readers of this book will likely interact with a humanoid robot before the decade ends—perhaps in a store, a clinic, an airport, or even their own living room. Others may find their jobs transformed by humanoid automation, their aging parents assisted by robotic caregivers, or their children growing up with humanoid companions as ordinary as the family dog once was.

Yet despite this approaching reality, most people remain unclear about what these machines actually are and how they work. This chapter provides that foundation—not with dense technical jargon or engineering diagrams—but with clear, accessible explanations. By the end of this chapter, readers will understand the essential distinction between AI and robotics, what makes a robot specifically "humanoid," why that human form carries both practical advantages and profound psychological implications, and where the technology stands today—capabilities, limitations, and all. With that clarity established, we can move forward together into the deeper questions of how humanoid robots will reshape our world and what we must do to ensure they serve human flourishing rather than diminish it.

Distinguishing Humanoid Robots from AI, Automation, and Other Robotics: What Makes a Robot "Humanoid"?

The confusion surrounding humanoid robots begins with language itself. When someone mentions "AI," they might mean the voice assistant on their phone, the algorithm recommending movies in their Gmail account, or the chatbot answering their customer service questions online. When they say "robot," they could be referring to the automated vacuum cleaning their floors, the surgical device assisting in an operating room, or the massive mechanical arms welding cars in a factory. These terms have become so interchangeable in everyday conversation that their distinct meanings have blurred into the vague definition "smart machines doing things."

Understanding what makes a robot specifically "humanoid" requires untangling this linguistic knot and recognizing that artificial intelligence, automation, robotics, and humanoid robotics represent different (though often overlapping) categories of technology. Each serves different purposes, operates under different constraints, and impacts daily life in fundamentally different ways.

Artificial intelligence refers to software and algorithms that enable machines to perform tasks typically requiring human intelligence, like learning from experience, recognizing patterns, making decisions, understanding language, and solving problems. AI exists purely as code and data without a physical form. The system that predicts what someone might want to buy online, the program that translates from one language to another, or the algorithm that detects fraud in financial transactions are all examples of AI operating entirely in the digital realm. AI powers countless systems people interact with daily, yet most of these interactions happen through screens, speakers, or behind-the-scenes processes invisible to the end users.

Automation describes systems designed to perform specific tasks with minimal human intervention, typically following predetermined rules and sequences. The thermostat that adjusts home temperature based on a programmed schedule, the assembly line that moves parts through different manufacturing stages, or the irrigation system that waters farm crops at set intervals are all automated systems that execute defined functions reliably and repeatedly. Automation has transformed industries for decades, but these systems generally operate within fixed parameters and perform the same tasks the same way unless reprogrammed by humans.

Robotics brings physical form to automated processes, creating machines that can move, manipulate objects, and interact with the physical world. This broad category encompasses everything from the robotic arms that assemble electronics to the wheeled devices that deliver medications in hospitals to the drones that survey farmland. What distinguishes robots from pure automation is their ability to sense their environment and respond with physical action. A factory robot might adjust its grip based on the weight of an object. A surgical

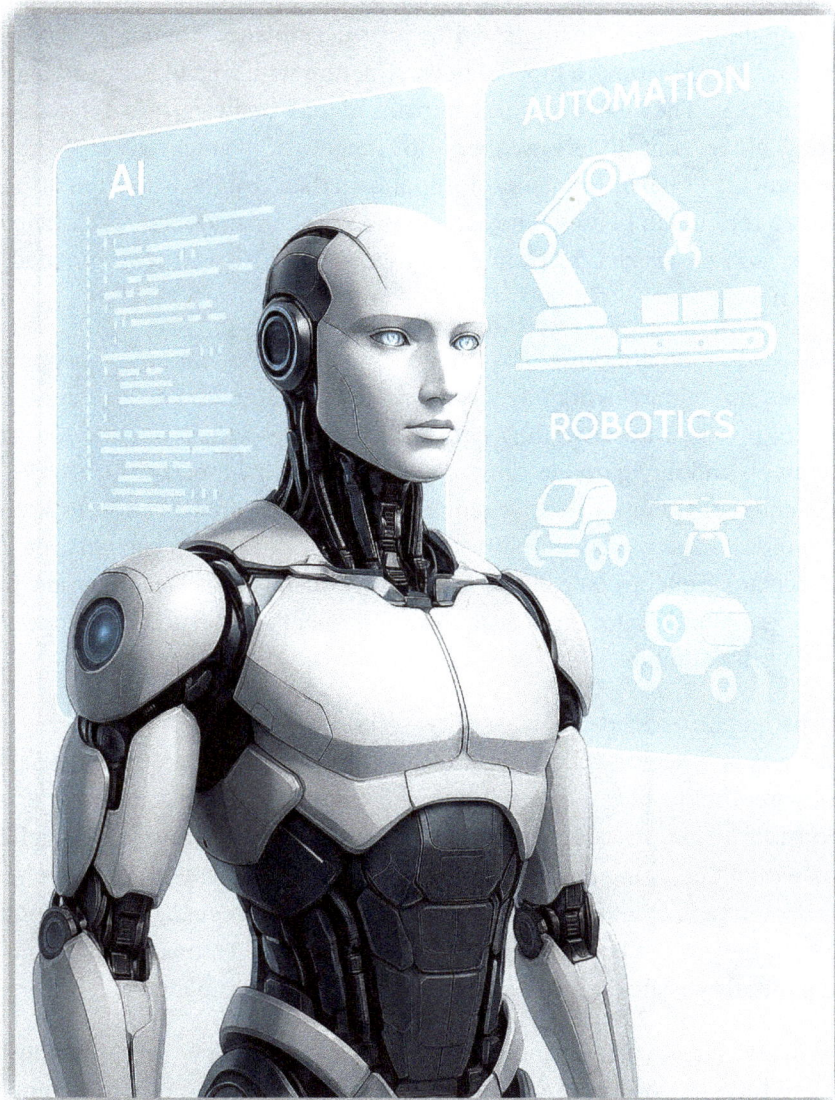

robot translates a surgeon's hand movements into precise incisions. These machines combine sensors, actuators, and control systems to perform physical tasks, but most are designed for specific environments and particular functions.

Humanoid robots represent a distinct subset of robotics, defined not just by their capabilities but by their deliberate resemblance to human form. A humanoid robot features a bipedal body structure with a head, torso, two arms, and two legs. They have a build mirrors human anatomy not for aesthetic reasons alone, but to allow them to functionally integrate into human environments.[3][5] Their humanoid form allows these robots to navigate spaces designed for human bodies. They can climb stairs, open doors, reach items on shelves, and sit in chairs, Humanoid robots can also use tools created for human hands.[1]

This physical similarity provides valuable practical implications. A warehouse designed for human workers doesn't need reconstruction to accommodate humanoid robots, because they can walk the same aisles, operate the same equipment, and navigate the same obstacles as their human counterparts.[4] A home built for a family doesn't require architectural modification to allow a humanoid assistant to move between rooms, reach kitchen cabinets, or help someone up from a chair. This adaptability distinguishes humanoids from specialized robots that require custom environments, dedicated workspaces, or specific infrastructure.[4]

The distinction extends to versatility. While a robotic arm might excel at welding or a wheeled robot might efficiently transport materials, humanoid robots are engineered as general-purpose machines capable of learning and performing varied tasks across multiple domains.[4] They can pick up delicate items without crushing them, figure out how to manipulate unfamiliar objects and adjust to unexpected obstacles, and transition between different activities. One hour they might be stacking boxes, then they might stop to assist someone with a mobility issue, before moving on to prepare a simple meal.[4]

What truly defines a humanoid robot is the convergence of a human-like physical form, AI-powered adaptability, sensory perception for environmental awareness, all of which provide a seamless integration into spaces and routines built around human needs and capabilities.[2][4] When these elements combine, the result is a machine fundamentally different from the AI assistant with no body, the automation system following fixed rules, or the specialized robot optimized for a single task. The humanoid robot represents technology shaped to fit into a human life rather than requiring human life to reshape around technology—a distinction that will determine how, where, and whether these machines can truly serve the people who need them most.[4]

Why Human Form Matters: The Psychology and Practicality of Building Robots That Look Like Us

The decision to build robots in human form might seem obvious at first glance. After all, humans naturally gravitate toward familiar shapes and faces. Yet the choice to invest billions of dollars and decades of engineering into replicating the human body rather than designing more efficient specialized machines reflects a convergence of profound psychological truths and hard practical realities that will determine whether these robots can actually serve the people who need them most.

The practical case for humanoid design begins with a simple observation: the modern physical world has been constructed around human proportions and capabilities. Doorways measure roughly three feet wide and seven feet tall because that accommodates a human body. Stairs are built in increments that human legs can comfortably climb. Kitchen counters sit at heights reachable by human arms. Light switches, doorknobs, steering wheels, shopping carts, hospital beds, and thousands of other everyday objects exist in configurations optimized for creatures with two hands, two legs, and a height between five and six feet. A robot designed to operate in this human-built environment faces a fundamental choice: adapt to the world as it exists or require the world to adapt to the robot.

Specialized robots like wheeled devices navigating warehouses, robotic arms assembling electronics, and automated systems performing surgery excel at specific tasks within carefully controlled environments. A factory floor can be redesigned around robotic arms. A warehouse can install specialized tracks and charging stations. But homes, small-town clinics, family farms, and the countless other spaces where people actually live out their daily lives cannot be rebuilt from scratch to accommodate machines. The elderly woman who needs help in her century-old farmhouse, the rural hospital operating in a converted Victorian building, the small business occupying a historic downtown storefront—these environments demand robots that can navigate human spaces, not spaces that can be reengineered for robots.

This is where the humanoid form delivers its most significant practical advantage.[4] A bipedal robot can climb the stairs in a split-level home that a wheeled device cannot navigate. Humanoid hands can turn the round doorknobs still common in older buildings, operate the manual controls on existing equipment, and pick up objects of varying sizes and fragility without requiring specialized grippers for each task.[3] The same humanoid robot that assists someone in a cramped apartment can work in a spacious suburban home, move between the tight aisles of a small-town grocery store, and navigate the uneven terrain of a rural property—all without environmental modification.[4]

Research into human-robot interaction reveals another reason why form matters: people respond to humanoid robots with greater intuitive understanding and trust than to machines that look purely mechanical. When a robot possesses a "head" that turns toward a person when they are speaking, humans instinctively interpret that as attention and engagement. When robotic arms reach and gesture, people know what to expect because they are mimicking their human arms.[2][6] This psychological comfort that humanoid robots provide makes collaboration easy and more natural.

The humanoid form also addresses a challenge particularly relevant in the context of caregiving: maintaining dignity and normalcy during vulnerable moments. An elderly person being assisted by a humanoid robot may feel the interaction is more familiar and less clinical than help from a faceless specialized medical device. A child learning alongside a humanoid tutor may engage and react more naturally than with a screen-based system. The human-like form doesn't replace genuine human connection, but it can make technological assistance feel less alienating and more a part of ordinary life.

Yet this psychological element of humanoid robots carries risks that cannot be ignored. The same human-like appearance that facilitates trust can also enable manipulation or create inappropriate emotional attachments. These are valid

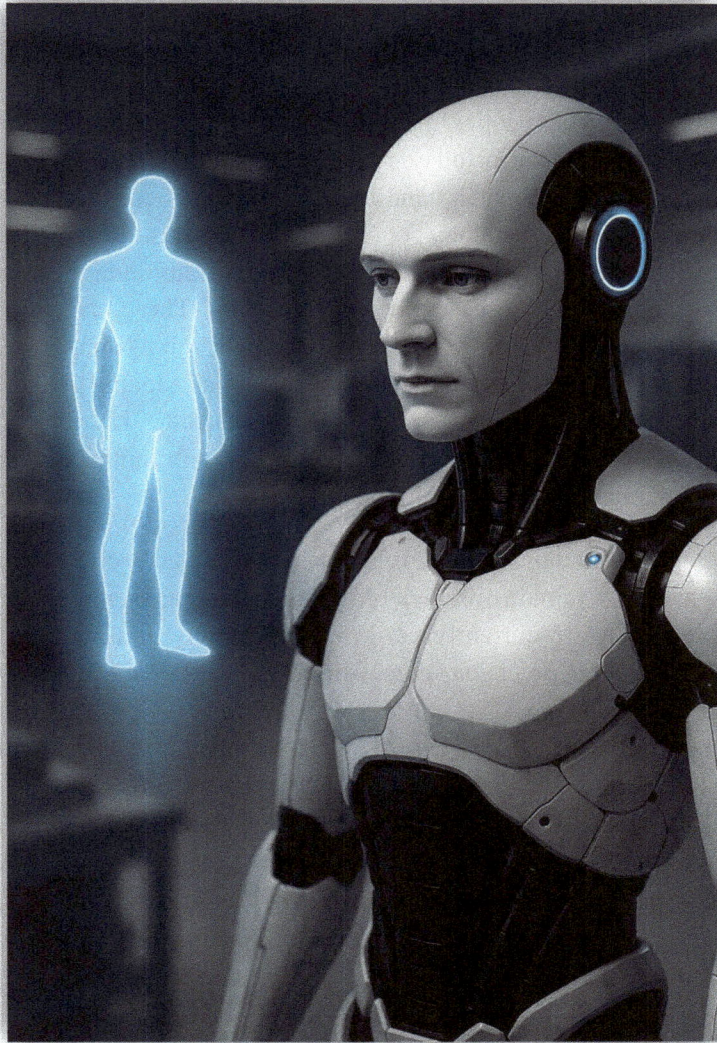

concerns that will require the setting of careful ethical boundaries as these machines are allowed to enter more intimate human spaces. The practical advantages of humanoid design come bundled with profound questions about what it means when the help we receive increasingly comes from entities that look like us but fundamentally are not. Ideally, communities must address these questions before humanoid robots become commonplace rather than after.

The Current Reality: What Today's Humanoid Robots Can (and Cannot) Actually Do

The gap between what people imagine humanoid robots can do and what they can actually accomplish in today's real-world settings represents one of the most consequential misunderstandings shaping public expectations today. Demonstrations at technology conferences showcase robots performing impressive feats—folding laundry, assembling automotive components, navigating obstacle courses with fluid motion. These carefully curated presentations create the impression that humanoid robots stand ready to step seamlessly into homes, hospitals, and workplaces as capable general-purpose assistants. The reality proves far more nuanced. There are genuine achievements in specific areas, but there are also substantial limitations that will constrain mass deployment for years to come.

Contemporary humanoid robots excel at tasks within controlled environments where variables remain predictable and parameters stay well-defined. Humanoid robots are currently performing assembly and material transport with sub-millimeter precision, demonstrating that they can reliably execute specific industrial workflows.[7] Tesla's Optimus Gen 2, standing 173 centimeters tall with 40 degrees of freedom, handles delicate objects and performs complex manipulation tasks in structured settings.[7] Boston Dynamics' Electric Atlas showcases remarkable dynamic capability to jump, spin, and manipulate tools. It has been specifically engineered for manufacturing environments where its bipedal motion navigates constrained spaces while human workers can focus on tasks with less physical strain and risk of injury.[7]

These robots have sophisticated sensory systems that enable them to autonomously navigate through semi-structured spaces.[7] Advanced vision systems combined with large language models allow them to understand tasks assigned via voice or visual cues, while cameras, LiDAR, force sensors, and tactile arrays provide real-time environmental feedback as they navigate their domain.[7] The infusion of generative AI represents perhaps the most significant recent advance, enabling robots to understand voice commands, ask clarifying questions, and respond conversationally.[7] Researchers project that high-level reasoning and spatial awareness capabilities may surpass human performance for some specific tasks within two to three years.[8]

Yet beneath these impressive capabilities lie fundamental limitations that prevent humanoid robots from functioning as the versatile assistants many envision. Core technologies remain below human capabilities in physical manipulation, particularly for tasks requiring nuanced force control or adaptation to unexpected variations.[8] A robot trained to fold specific garment types may fail when confronted with unfamiliar fabrics or configurations. The dexterity gap becomes especially apparent in genuinely unstructured environments such as the messy reality of human homes with pets and clutter, rural properties with uneven terrain, or public spaces with unpredictable obstacles and variable lighting conditions.[8]

Today's humanoid robots struggle with adapting beyond their training scenarios. While language models function across linguistic domains with relative ease, robots cannot readily transfer learned skills from one physical context to another. A robot performing assembly tasks in one factory configuration may require substantial retraining when deployed in a different facility. This limitation reflects a fundamental gap between demonstrated capability and operational reliability and is the difference between what a robot can do one time under ideal conditions and what it can accomplish consistently multiple times across varied real-world situations.

Battery technology imposes additional constraints. While Figure 02's battery lasts over 20 hours and Tesla Optimus operates through a full workday, these capabilities depend on the efficient execution of their designated tasks.[7] More complex or varied work demands higher energy consumption, and extended periods of cognitively demanding activity use more energy, as well. Energy constraints become particularly relevant for applications in rural areas or caregiving contexts where continuous operation without frequent charging access may be necessary.

Perhaps most significantly, contemporary humanoid robots are confined to controlled or semi-structured settings.[8] Manufacturing plants, warehouses, and designated service areas provide the predictable conditions these machines require. The genuinely open real world of crowded public streets, homes filled with infinite variability, and caregiving situations demanding nuanced social interpretation, remains beyond their operational capacity. Within five years, improved dexterity and battery modules could support semi-structured service settings like hotel rooms or hospital supply rooms.[8] The breakthroughs required to enable elder care assistance, light construction, or handling remote materials remains a decade away or more.[8]

Understanding this reality and recognizing both the current capabilities and limitations leads to more realistic expectations about when and how humanoid robots will actually enter daily life. The journey from confusion to clarity regarding humanoid robots begins with understanding what these machines actually are—not the science fiction fantasies that have shaped expectations for generations, but the tangible technology now moving from research laboratories into real-world applications. A humanoid robot is neither pure artificial intelligence existing only as code nor a specialized machine optimized for a single task. It represents something fundamentally different: a physical embodiment designed to navigate human spaces, manipulate human tools, and integrate into environments built around human proportions and capabilities without requiring those environments to be altered to support them.

This distinction matters profoundly for the people who will actually encounter these machines in the coming years. The elderly farmer who needs help maintaining property that has been in the family for generations cannot redesign a century-old barn to accommodate wheeled robots or install specialized infrastructure for automated systems. The small-town clinic operating in a converted historic building cannot rebuild its layout around robotic equipment that requires custom environments. The family caring for an aging parent in a modest home filled with a lifetime of belongings cannot transform their living space into a controlled laboratory setting. These real-world contexts demand robots that adapt to human life as it actually exists—messy, unpredictable, infinitely varied, and deeply rooted in spaces designed long

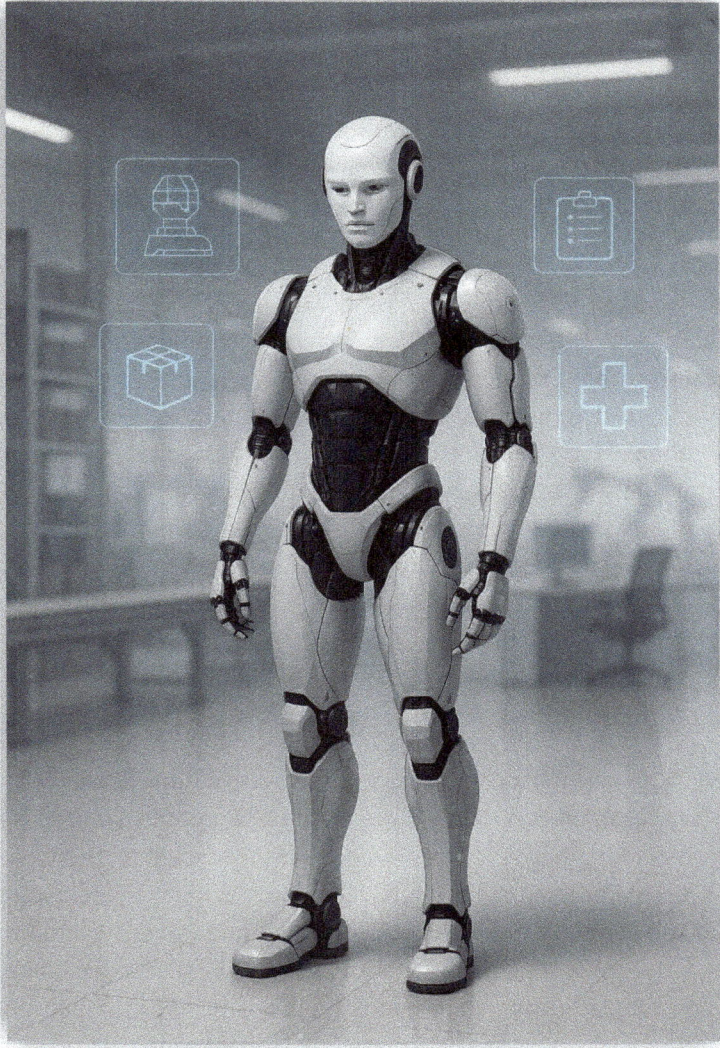

before anyone imagined machines walking among us.

The humanoid form delivers this adaptability not through technological magic but through deliberate design choices grounded in both practical necessity and psychological reality. When robots can climb stairs, turn doorknobs, reach shelves, and manipulate objects with human-like hands, they gain access to the vast infrastructure of human civilization without requiring that infrastructure to change. When they possess heads that turn toward who is speaking and arms that gesture with recognizable motions, they open the door to interact with for people who don't know anything about computers or technology. This ideal convergence of form and function explains why engineers have invested decades pursuing the considerable challenges of creating bipedal locomotion and dexterous manipulation rather than settling for more efficient specialized designs.

Yet understanding what humanoid robots are requires equal clarity about what they are not—at least not yet. The current generation excels in controlled manufacturing environments, performs specific tasks with impressive precision, and demonstrates remarkable capabilities in carefully curated demonstrations. These achievements will transform certain industries within years. But the versatile, adaptable, general-purpose assistants capable of navigating genuinely unstructured environments—homes filled with pets and clutter, rural properties with unpredictable terrain, caregiving situations demanding nuanced social interpretation—remain years or even decades from deployment. The gap between demonstration capability and operational reliability, between what works once under ideal conditions and what functions consistently across varied real-world situations, still yawns wide.

This honest assessment serves not to diminish the significance of humanoid robotics but to ground expectations in reality. The robots entering warehouses and factories today will teach invaluable lessons about safety, reliability, and human-robot collaboration that will shape the machines that eventually arrive in homes and communities. The technological limitations constraining current systems will drive innovations to overcome them. Understanding where humanoid robots truly stand today—their genuine capabilities, their substantial constraints, and the timeline for meaningful advancement— provides the foundation necessary for the deeper questions ahead: How will these machines transform work and caregiving? What ethical boundaries must guide their development? How can communities prepare for a future where robots walk among us as daily companions in the most intimate spaces of our human life?

CHAPTER 2. THE RISE OF HUMANOID ROBOTS: FROM SCIENCE FICTION TO YOUR FRONT DOOR

The first time most people encounter a humanoid robot in person (not in a video or news article, but standing in the same room) they report feeling something unexpected—not quite fear, not quite wonder, but a disorienting sense that the future has arrived without asking permission. That feeling matters, because understanding how we arrived here helps us make sense of where we're headed and why the pace of change suddenly feels so breathtaking.

For decades, humanoid robots lived primarily in our imaginations. They walked through science fiction novels, served drinks on *The Jetsons*, threatened humanity in *Terminator* films, and gently cared for children in futuristic stories like *A.I. Artificial Intelligence*. These fictional robots shaped our expectations in powerful ways, creating both unrealistic hopes (where's my robot butler?) and exaggerated fears (will they turn against us?). The gap between what we imagined and what engineers could actually build remained enormous for so long that many people stopped believing humanoid robots would ever truly arrive. The technical challenges of balance, dexterity, power management, and a sophisticated artificial intelligence , all proved far more difficult than mid-century futurists predicted.

Yet something fundamental shifted in the past decade. The same forces that brought us smartphones, self-driving car prototypes, and AI that can write essays and generate art have finally converged to make humanoid robots not just possible, but practical. Advances in machine learning allow robots to learn from experience rather than requiring programmers to anticipate every scenario. Improvements in battery technology, lightweight materials, and sensor arrays have made human-scale robots more capable and affordable. Computing power that once filled entire rooms now fits in a robot's torso. Most importantly, the economic and social pressures driving humanoid robot development—aging populations, caregiver shortages, dangerous working conditions, labor costs—have reached a tipping point where investing billions in humanoid robotics suddenly makes business sense.

This chapter traces that journey from imagination to implementation, not as a dry technical timeline, but as a story about human ambition, setbacks, breakthroughs, and the specific innovations that finally closed the gap between science fiction and science fact. Understanding this history matters because it helps separate realistic expectations from hype, explains why certain companies and countries are leading the charge, and reveals what problems humanoid robots are actually designed to solve (which isn't always what movies taught us to expect).

The robots emerging from research labs and entering pilot programs in warehouses, hospitals, and homes today represent the culmination of nearly a century of dreaming and decades of painstaking engineering work. They're not the sentient beings of fiction, but they're far more capable than the clunky prototypes that stumbled through demonstrations just ten years ago. By understanding how we got here—what took so long, what finally changed, and who's building these machines now—readers can better assess what's coming next and what it will mean for their own lives, jobs, families, and communities. The humanoid revolution didn't happen overnight, but it is happening now.

From Rosie to Reality: How Science Fiction Shaped Our Expectations (and Why Progress Took Longer Than We Thought)

The cultural blueprint for humanoid robots emerged not from engineering laboratories but from television screens, movie theaters, and paperback novels. When *The Jetsons* premiered in 1962, Rosie the Robot became America's template for what a household robot should be: helpful, conversational, capable of complex tasks, and possessing enough personality to feel like a member of the family rather than an appliance.[9][10] This animated maid, with her sassy demeanor and tireless work ethic, shaped expectations for generations of viewers who grew up assuming that by the time they reached adulthood, they too would have a Rosie managing their households.

Science fiction didn't just entertain—it created a promise. From Isaac Asimov's thoughtful robots governed by ethical laws to *Star Wars* ' endearing C-3PO and from *Star Trek* 's loyal android Data to the menacing Terminator, fictional robots established the boundaries of what seemed possible and desirable. These stories taught audiences that humanoid robots would walk smoothly, understand complex instructions, navigate unpredictable environments, and interact naturally with humans. The cultural consensus became clear: humanoid robots weren't a question of "if" but "when," and that "when" felt perpetually just around the corner.[10]

The reality, however, proved far more stubborn than the fiction suggested. The technical challenges that engineers faced (and continue to face) were missing from the smooth animations and carefully choreographed film sequences that made robot movement look effortless. Creating a machine that could maintain balance while walking across uneven surfaces required solving problems in dynamics, control theory, and sensor integration that took decades of research. Teaching a robot to grasp a fragile object without crushing it, or to understand that "clean the kitchen" meant washing dishes, wiping counters, and sweeping floors rather than simply standing motionless, demanded advances in artificial intelligence, computer vision, and manipulation that remained out of reach for generations.

The gap between expectation and capability created what some call "the Rosie disappointment"—a recurring pattern where companies would announce a huge breakthrough in household robots, generate enormous excitement, and then deliver products that could barely vacuum a room without getting stuck. The promises of the 1970s and 1980s, when various firms claimed domestic robots were imminent, collapsed under the actual technical limitations.[10] Battery technology couldn't power a human-sized robot for more than brief demonstrations. Computer processors lacked the speed to process sensor data in real time. Machine learning algorithms that could enable robots to adapt to new situations simply didn't exist yet.

What science fiction had made look simple such as Rosie effortlessly carrying a tray of dishes while conversing with the Jetson family, turned out to require solving some of the hardest problems in engineering and computer science simultaneously. Walking requires constant micro-adjustments to maintain balance by processing input from dozens of sensors hundreds of times per second. Manipulating objects demands understanding their properties, predicting how they'll respond to force, and adjusting grip strength in milliseconds. Understanding human speech in noisy, real-world environments with varied accents and imprecise phrasing remained largely unsolved until recent advances in neural networks and natural language processing.

The cultural impact of this gap matters because it shaped how people responded to actual robotics progress. When researchers demonstrated robots that could walk slowly across flat surfaces or pick up specifically shaped objects in controlled laboratory settings, the public reaction was often disappointment rather than appreciation. After all, Rosie could do so much more, and she was invented in 1962. This lack of understanding about the genuine difficulty involved in creating humanoid robots sometimes translated into skepticism about whether humanoid robots would ever truly arrive. Overpromising led to underdelivering, which in turn bred cynicism.

Yet the science fiction vision, for all its technical naivety, captured something important: the human desire for machines designed for human needs that could work alongside people in human spaces. That vision, refined by decades of hard-won engineering progress, is finally approaching reality—not because the problems became easier, but because researchers developed the tools, algorithms, and materials needed to solve them.

The Breakthrough Decade: What Changed to Make Humanoid Robots Suddenly Possible

For years, roboticists could build machines that excelled at single tasks like welding car frames, vacuuming floors, or even performing delicate surgeries. But creating a robot that could walk across a room, pick up a dropped medication bottle, and hand it to an elderly person remained impossibly complex. Then, in the span of roughly a decade, everything changed. The period from approximately 2010 to 2020 witnessed a convergence of technological breakthroughs that transformed humanoid robots from expensive laboratory curiosities into machines approaching practical usefulness. Understanding what shifted during these years helps explain why humanoid robots are arriving now rather than remaining perpetually "twenty years away."

The single most transformative development was the deep learning revolution that began in 2012. That year, computer scientist Geoffrey Hinton and his team at the University of Toronto achieved something remarkable: they created a visual recognition system that could identify and sort over a million images with an error rate of just 15.3 percent—a full ten-point improvement over previous methods.[13] This breakthrough mattered enormously for humanoid robotics because robots operating in human environments need to understand what they're seeing. A robot caring for an aging parent needs to distinguish between a water glass and a medication bottle, recognize when someone has fallen, and navigate around furniture that gets moved. Deep learning made this level of visual understanding possible for the first time, and when the technology became accessible to researchers and companies beyond elite laboratories, it created innovation across the field.[13]

Simultaneously, sensor technology made dramatic leaps forward. Driven by autonomous vehicle research, the development of advanced cameras, depth sensors, and Lidar systems gave robots the ability to perceive their three-dimensional environment with unprecedented precision.[13] Boston Dynamics' Atlas robot introduced in 2013 demonstrated how these sensors could be integrated into a humanoid platform, enabling the machine to maintain balance when pushed, navigate obstacles, and adjust its movements in real time.[11][12] These weren't incremental improvements; they represented fundamental shifts in what robots could perceive and respond to in unpredictable real-world settings.

Battery technology and motor innovation provided equally critical advances. For decades, humanoid robots were tethered to power sources or could operate for only brief periods before exhausting their batteries. Improvements in battery energy density driven partly by smartphone and electric vehicle development finally gave humanoid robots the endurance needed for practical tasks. Simultaneously, innovations in motor design, particularly the development of more compact and efficient actuators, allowed engineers to build robots with the strength and precision needed for delicate manipulation while maintaining human-like proportions and weight distribution.[11]

The breakthrough decade also benefited from exponential increases in computing power that could fit inside a robot's body. The processors required to analyze sensor data, run deep learning algorithms, and control dozens of motors simultaneously and all in real-time became small enough and power-efficient enough to be embedded in mobile platforms. This meant humanoid robots could finally process information and make decisions without requiring constant connection to remote supercomputers.

Perhaps most importantly, these separate streams of innovation converged. A humanoid robot needs sophisticated AI, advanced sensors, capable motors, sufficient power, and powerful onboard computing all working together seamlessly. During the breakthrough decade, these technologies matured simultaneously and became affordable enough that not just wealthy corporations could attempt humanoid development. By 2020, the fundamental technological barriers had been overcome, shifting the question from "Can we build functional humanoid robots?" to "What should we build them to do, and how do we ensure they serve human needs responsibly?"

The robots emerging from laboratories today represent the culmination of this convergence. They are machines that can finally begin to approach the capabilities science fiction promised, though still with significant limitations.

Who's Building Them and Why Now: The Companies, Research Labs, and Real-World Needs Driving the Humanoid Revolution

The companies and research institutions racing to build humanoid robots today represent an unusual convergence of established industrial giants, ambitious startups, and technology firms better known for electric vehicles or artificial intelligence than for robotics. Understanding who's building these machines and why they've chosen this particular moment to invest billions of dollars in humanoid development reveals as much about the problems these robots are meant to solve as it does about the technology itself. The humanoid revolution isn't happening because engineers suddenly figured out how to make robots walk; it's happening because real-world pressures—labor shortages, aging populations, dangerous working conditions, and economic forces—have reached a tipping point where humanoid robots finally make business sense.

Tesla's entry into humanoid robotics with its Optimus robot surprised many observers who associated the company exclusively with electric vehicles, but the move reflects strategic logic.[14] Tesla already possessed the AI systems, computer vision technology, and manufacturing scale developed for autonomous driving, and these are all capabilities that translate directly to humanoid robotics.[14] The company has outlined plans to scale production to one million units annually by 2030, a manufacturing target that would have seemed absurd just five years ago but now represents the kind of ambition driving the entire industry.[14] Tesla's involvement matters not just because of its technical capabilities, but because it signals that humanoid robots have moved from research curiosities to products that major corporations believe will generate substantial revenue.

Boston Dynamics, the veteran robotics company whose viral videos of backflipping robots captivated millions, recently transitioned its Atlas

humanoid from research demonstrations to commercial applications, beginning with part sequencing in automotive manufacturing.[17] This shift from "look what our robot can do" to "here's the specific problem it solves for paying customers" marks a fundamental change in how even the most technically sophisticated robotics companies approach humanoid development. Marc Raibert, Boston Dynamics' founder, acknowledged that the broader industry momentum—including Tesla's work—has validated the commercial potential that his company spent decades developing.[17]

Agility Robotics achieved a historic milestone when its Digit robot became the first humanoid to perform paid work in a real production environment.[17] Under a multiyear agreement with GXO Logistics, Digit moves boxes and places them onto conveyor belts at Spanx manufacturing facilities. As of 2025, the robot had moved over 100,000 totes.[17] This isn't a demonstration or pilot program, it's sustained commercial operation where a humanoid robot performs repetitive, physically demanding work that human workers increasingly avoid. The company has expanded deployments to Amazon and other logistics operations, demonstrating that customers will pay for humanoid labor when it addresses genuine operational challenges.[17]

Newer entrants like Figure AI, backed by substantial venture capital, and 1X, supported by OpenAI, represent a wave of startups betting that general-purpose humanoid robots will transform multiple industries simultaneously.[15][17] Figure's leadership has stated that humanoid robots capable of useful work are "single-digit years away," a timeline that reflects both technological confidence and market urgency.[17] Sanctuary AI's Phoenix robot has already completed paid shifts in retail and e-commerce fulfillment, initially under remote operation but increasingly transitioning to autonomous execution and demonstrating the progression from human-controlled to independent robot work.[16]

The "why now" question has a clear answer: labor markets in developed economies face unprecedented pressures from aging populations, declining birth rates, and worker shortages in physically demanding roles. Manufacturing facilities, warehouses, and logistics operations struggle to attract workers for repetitive, dangerous, or monotonous tasks. Simultaneously, the technological barriers that prevented functional humanoid robots for decades—inadequate AI, insufficient battery life, limited sensors, weak computing power—have finally been overcome through the convergence of breakthroughs in machine learning, computer vision, and materials science.[14][15] Companies aren't building humanoid robots because the technology is cool; they're building them because customers desperately need solutions to labor challenges that traditional automation cannot address, and the technology has finally matured enough to deliver those solutions profitably.

The journey from Rosie the Robot to the humanoid machines now entering warehouses and pilot programs spans more than six decades, but the most dramatic transformation has occurred in just the past ten years. Understanding this acceleration matters because it reveals something essential about the moment we're living through. Humanoid robots aren't arriving because engineers suddenly became smarter or more ambitious, but because multiple technological breakthroughs converged with urgent real-world needs that could no longer be ignored. The labor shortages straining hospitals, warehouses, and farms aren't abstract future problems—they're present-day crises that have made humanoid development not just technically possible but economically necessary.

The companies investing billions in humanoid robotics today—from Tesla's manufacturing ambitions to Boston Dynamics' commercial pivot to startups like Agility Robotics—represent a fundamental shift from research curiosity to practical tool. When a humanoid robot moves its hundred-thousandth box in a real production facility, when logistics companies sign multiyear contracts for robot workers, when automotive manufacturers integrate humanoids into assembly lines, then these aren't demonstrations of what might someday be possible. They're evidence that the future science fiction promised has quietly begun arriving.

Yet this history also teaches caution about separating genuine capability from marketing hype. The gap between what fictional robots could do effortlessly and what real machines struggled to accomplish created decades of disappointment and skepticism that still shapes how people respond to robotics announcements today. The robots emerging from laboratories in 2025 remain far more limited than Rosie, C-3PO, or the Terminator. Today's robots can't truly understand context the way humans do, they struggle with tasks that require common sense reasoning, and they operate reliably only in relatively controlled environments. The breakthrough decade solved enormous technical problems, but it didn't solve all of them.

What has changed irrevocably is the trajectory. The technological foundations of deep learning, advanced sensors, capable batteries, and powerful embedded computing are now mature enough that continued improvement seems inevitable rather than speculative. The economic pressures driving development—aging populations in wealthy nations, worker shortages in physically demanding industries, the relentless push to reduce labor costs— show no signs of diminishing. The companies and research institutions building humanoid robots today possess resources, expertise, and market validation that previous generations of roboticists could only imagine.

For readers wondering when humanoid robots will truly affect their own lives, workplaces, and communities, this history provides a framework for realistic expectations. The robots arriving in the next few years will likely appear first in commercial and industrial settings where controlled environments, specific tasks, and economic pressures make deployment practical. They'll gradually improve through accumulated experience and continued innovation, much as smartphones evolved from expensive novelties to indispensable tools over a decade. The timeline for humanoid robots becoming common in homes, hospitals, and small towns remains uncertain and is likely measured in years rather than decades.

The science fiction that shaped our expectations wasn't wrong about the destination, only about how long and difficult the journey would prove. We're finally approaching the world those stories imagined, which means the questions shift from "Will humanoid robots arrive?" to "How do we ensure they serve human flourishing rather than undermine it?" The next chapters explore those questions in the specific contexts where humanoid robots will matter most—in our work, our homes, helping care for aging loved ones, and overseeing our children's development—because understanding the technology's history is only the beginning of preparing for its impact

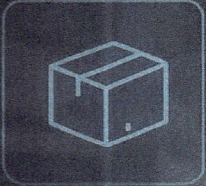

CHAPTER 3. HOW ROBOTS WILL CHANGE THE WORLD OF WORK: JOBS, DISPLACEMENT, AND ADAPTATION

The conversation about humanoid robots and employment tends to split into two camps: those who insist robots will steal every job and leave humanity obsolete, and those who promise technology always creates more opportunities than it destroys. Both narratives oversimplify a transformation that will be far messier, more uneven, and more human than either extreme suggests. The truth is that humanoid robots will indeed change the landscape of work profoundly. They will eliminate some jobs entirely, transform others beyond recognition, and create roles we can barely imagine today, but what that means for workers, families, and communities depends enormously on choices we make right now about preparation, education, policy, and values.

This chapter examines those changes with the honesty they deserve. Unlike previous waves of automation that primarily affected manufacturing assembly lines or replaced human labor with specialized machines bolted to factory floors, humanoid robots represent something fundamentally different: mobile, adaptable machines that can navigate human spaces, manipulate objects designed for human hands, and potentially perform any physical task a human body can accomplish. That versatility is precisely what makes them both promising and threatening. They're not limited to repetitive motions in controlled environments, but can theoretically work in uncontrolled environments like warehouses, retail stores, hospitals, hotels, farms, and homes. The question isn't whether this will disrupt employment, as it surely will, but which jobs will disappear, which will evolve, who will bear the costs of transition, and whether we can build an economy where human workers thrive alongside their robotic counterparts rather than compete desperately against them.

The stakes are particularly high for workers without college degrees, for rural communities already struggling with economic decline, for older workers facing the prospect of retraining in their fifties and sixties, and for families living paycheck to paycheck without safety nets to cushion disruption. These aren't abstract economic statistics—they're neighbors, relatives, and fellow community members whose livelihoods and dignity hang in the balance. At the same time, humanoid robots offer genuine solutions to real problems such as dangerous jobs that break human bodies, labor shortages in essential industries, physically demanding work that aging populations can no longer sustain, and tasks that keep humans from higher-value activities requiring creativity, empathy, and complex judgment.

This chapter will explore which jobs face the greatest automation risk and why, examining not just what robots can technically do but what economic and social factors will drive their adoption. It will look beyond simple replacement scenarios to consider hybrid roles where humans and robots collaborate, each contributing what they do best. Most importantly, it will offer practical, actionable guidance for workers and communities facing this transition—not empty reassurances that everything will work out fine, but real strategies for building adaptability, identifying durable skills, and advocating for policies that ensure the benefits of automation are shared rather than concentrated. The robot revolution in the workplace is coming whether we feel ready or not, but we still have a say in shaping what that revolution looks like and who it serves.

The Honest Assessment: Which Jobs Are Most Vulnerable to Humanoid Automation (and Which Aren't)

Understanding which jobs face the greatest risk from humanoid automation requires moving beyond simplistic predictions about robots "taking all our jobs" toward a more nuanced examination of tasks, skills, and economic realities. The research reveals a critical distinction: humanoid robots will not eliminate entire occupations but instead will automate specific tasks within jobs.[18] This task-based distinction means that even positions facing significant automation may evolve rather than disappear entirely, while others will remain largely untouched by robotic capabilities for the foreseeable future.

The jobs facing the most immediate and substantial threat share common characteristics. Routine manual work—tasks that follow predictable patterns in controlled environments—represents the lowest-hanging fruit for humanoid automation.[18] These routine tasks include manufacturing assembly positions, warehouse operations involving selecting and packing, and basic food preparation roles involve repetitive physical movements that humanoid robots can replicate. Research tracking actual robot deployment in the United States found that for every robot added per 1,000 workers, employment declined by approximately 0.2 percentage points, translating to roughly 400,000 jobs affected as of 2025.[19] Workers in these positions often lack college degrees and work in America's industrial regions, where a single robot's introduction can displace multiple workers within the same geographic area.[18][19]

Retail and basic customer service positions face growing vulnerability as humanoid robots become more sophisticated. Tasks that don't require complex problem-solving or nuanced human judgment like stocking shelves, inventory management, checkout operations, and standardized customer interactions can be handled by robots that are able to navigate store environments and interact with customers for routine transactions. The threat extends beyond direct replacement: as manufacturing workers lose positions to automation, many transition into retail and service work, creating wage pressure even in jobs not directly automated.[19]

Administrative and clerical roles involving data entry, basic scheduling, routine processing, and standardized communication also face automation risk.[18] While not typically associated with humanoid robots, these positions illustrate how automation broadly threatens any work that follows established procedures without requiring creative adaptation or complex judgment.

Demographic patterns reveal that automation's burden falls unevenly. Research examining robot deployment between 1993 and 2014 found that men experienced employment reductions of 3.7 percentage points compared to 1.6 points for women, while non-White workers saw employment cut by 4.5 percentage points versus 1.8 points for White workers.[20] These disparities reflect occupational segregation, with vulnerable populations who are concentrated in the manufacturing and routine manual positions being the most susceptible to robotic replacement. Workers without college degrees face substantially greater negative impacts than those with higher education, as their jobs more often involve the routine physical tasks that robots can master.[18]

Conversely, certain categories of work remain substantially resistant to humanoid automation, at least with current and near-future technology. Jobs requiring complex problem-solving, strategic thinking, and creative innovation like management positions, research roles, and engineers, demand the ability to adapt to fluid situations that robots cannot yet replicate. The unpredictability and judgment required make these positions relatively secure.

Work involving genuine emotional intelligence and nuanced interpersonal interaction also remains largely protected. While humanoid robots can handle basic customer service scripts, positions requiring empathy, complex relationship-building, counseling, mentoring, and authentic human connection depend on qualities robots cannot authentically provide.[21] Healthcare professions, particularly those involving diagnosis, treatment decisions, and patient care requiring judgment, remain substantially resistant to automation despite robots' growing presence in medical settings in other roles.[21]

Skilled trades present an interesting case. Plumbing, electrical work, carpentry, and HVAC repair involve physical work in unpredictable environments where every job site presents unique challenges. The variability, environmental complexity, and on-the-fly problem-solving required keep these positions relatively safe from automation, even though robots are becoming more dexterous and mobile.

It turns out then that neither the dystopian vision of mass unemployment nor the utopian promise that automation always creates more jobs than it destroys is accurate. Instead, the evidence points toward a disruption impacting mainly workers already economically vulnerable, in regions already struggling, and performing tasks that are routine and predictable. The outcome of this impact depends not on technology alone, but on the choices communities and policymakers make about supporting workers through transition, investing in education and retraining, and ensuring automation's benefits reach beyond corporate balance sheets to the people whose livelihoods hang in the balance.

Beyond Replacement: How Human-Robot Collaboration Will Create Hybrid Roles and Augment Human Capabilities
The conversation about robots and work need not be framed as an either-or proposition where either humans keep their jobs or robots take them all. The emerging reality suggests something far more nuanced and potentially more hopeful: a workplace where humans and robots form partnerships that leverage the distinct strengths each brings to the table. Research demonstrates that human-robot teams achieve 85% higher productivity than those working alone[24], not because robots work faster, but because strategic collaboration allows each participant to contribute what they do best.

This collaborative model fundamentally changes how work gets organized. In manufacturing settings, humanoid robots might handle the physical assembly of products—the repetitive motions that strain human bodies and dull human minds—while human workers focus on quality control, process improvement, and customer interaction[25]. The robot doesn't steal the assembly worker's job;

it transforms that job into something requiring different skills, often higher-level cognitive functions that machines cannot replicate. The assembly worker becomes a quality assurance specialist, a process innovation analyst, or a robot operations coordinator—roles that didn't exist before but would emerge naturally from the partnership.

Healthcare offers particularly compelling examples of this augmentation model. Humanoid robots can monitor patients continuously, track vital signs, remind elderly residents to take medications, and provide companionship during long hospital nights[22]. These tasks consume enormous amounts of human nursing time that could then be redirected toward complex medical decision-making, emotional support for frightened patients, and the nuanced judgment calls that define excellent care. A nurse working alongside a humanoid assistant doesn't become obsolete; she becomes more effective, freed from routine monitoring to focus on the irreplaceable human elements of healing. The robot handles the predictable; the human handles the unpredictable.

In the logistics industry, Agility Robotics' Digit robot, deployed through a multi-year agreement with GXO Logistics, handles tasks like unloading trailers and managing inventory[22]—physically demanding work that often causes injuries and exhausts human workers. The humans in these facilities then shifted from lifting boxes to managing complex shipment scenarios, handling exceptions that require judgment calls, and coordinating operations that demand strategic thinking. The human work doesn't disappear; it evolves into something less physically punishing and more cognitively engaging.

This transformation also creates entirely new categories of employment. Workers who can program, troubleshoot, and optimize robotic systems like robot operators and trainers—represent roles that barely existed a decade ago but will become commonplace as humanoid robots themselves become more common[22]. Safety compliance officers who monitor both human and robotic operations by ensuring safe workplace conditions in mixed environments, fill a critical need emerging from the human-robot collaboration. Automation analysts evaluate which tasks suit robotic automation and which require human judgment, making strategic decisions that shape how work gets divided between flesh and machine.

The safety benefits alone justify serious consideration of human-robot collaboration. Humanoid robots can handle hazardous tasks—working with toxic chemicals, operating in extreme temperatures, performing repetitive motions that cause cumulative injuries—without risking human health and safety[22] [23]. Workers previously assigned to dangerous roles can transition to positions focused on safety management, risk mitigation, and emergency response coordination. The reduction in workplace injuries represents not just cost savings for employers but preserved health and dignity for workers who no longer sacrifice their bodies to earn a living.

Yet this optimistic vision requires acknowledging uncomfortable realities. The transition from routine manual work to higher-level cognitive roles demands substantial retraining, and not every displaced worker will successfully make that leap. Organizations must actually invest in workforce development and not just offer token training programs while quietly planning layoffs. Ideally, workers need to be involved in shaping how robots integrate into their workflows, not top-down mandates that treat employees as obstacles to efficiency[25]. The productivity gains from human-robot collaboration should benefit workers through better wages and working conditions, not be funneled back exclusively to shareholders while employees face stagnant pay and job insecurity.

Evidence suggests that human-robot collaboration, implemented thoughtfully and ethically, can create workplaces where humans focus on creativity, problem-solving, and interpersonal connection while robots handle the repetitive and dangerous[25]. That future isn't guaranteed as it requires deliberate choices about worker protection, fair compensation, and genuine investment in human capability. But it remains an achievable option, and far preferable to either wholesale displacement or refusing technological progress altogether.

Preparing for the Shift: Practical Strategies for Workers, Families, and Communities to Adapt and Thrive

The question facing workers today isn't whether humanoid robots will transform employment—it's how to position oneself, one's family, and one's community to navigate that transformation with dignity and opportunity intact. The window for proactive preparation is narrowing as humanoid robots approach cost parity with human labor[26] and deployment accelerates across manufacturing, logistics, healthcare, and service sectors. Yet preparation does not mean panic. Deliberate adaptation strategies at individual, household, and community levels can substantially mitigate disruption while positioning people to capture emerging opportunities that robot integration creates.

The foundation of individual resilience begins with honest self-assessment. Workers should evaluate which aspects of their current roles involve routine, predictable physical tasks and which require creativity, complex judgment, emotional intelligence, or adaptability to novel situations. A warehouse worker who spends most shifts moving boxes faces more vulnerability than one who troubleshoots logistics problems, coordinates with customers, or trains new employees. This assessment isn't about inducing anxiety but about creating clarity: understanding one's position relative to automation trends enables strategic rather than reactive responses.

Building on that clarity, workers can deliberately cultivate skills that complement rather than compete with robotic capabilities. Technical literacy—understanding how to work alongside robotic systems, basic programming concepts, and maintenance fundamentals—transforms workers from automation victims into automation partners. Manufacturing facilities deploying humanoid robots report that human employees who develop these competencies transition into roles like a robot operations coordinator, quality assurance specialist, and process improvement analyst roles.[25] These aren't abstract future jobs but roles being created right now in facilities where Digit robots unload trailers and Optimus prototypes handle assembly tasks.

Equally valuable are the distinctly human capabilities that remain beyond robotic reach: creative problem-solving, empathy and relationship-building, strategic thinking, and the ability to navigate ambiguous situations requiring judgment rather than procedure.[25] A home health aide who develops expertise in family counseling and care coordination becomes more valuable even as robots handle medication reminders and mobility assistance. A retail worker who cultivates customer relationship skills and community knowledge offers something humanoid robots cannot replicate, regardless of their dexterity or speech capabilities.

Families facing this transition should approach adaptation as a collective rather than individual challenge. Household diversification—ensuring multiple family members develop complementary skills across different sectors—reduces vulnerability to any single industry's disruption. Building emergency savings becomes particularly critical when job transitions may require months of retraining or relocation. Understanding available support systems such as unemployment insurance, retraining programs, community college resources, and apprenticeship opportunities can transform abstract safety nets into concrete lifelines when displacement occurs.

Communities, particularly those dependent on manufacturing or routine physical labor, face the most profound adaptation challenge. The evidence from previous automation waves shows that regions failing to diversify economically experience cascading decline: job losses reduce tax revenues, diminish public services, drive out younger residents, and create downward spirals difficult to reverse. Proactive communities are investing now in educational infrastructure by expanding community college technical programs, creating partnerships between schools and employers deploying robots, and building entrepreneurship support systems that help displaced workers start new ventures rather than simply seeking traditional employment.

The timeline for these preparations is compressed. Analysts project that physical jobs across multiple sectors face meaningful disruption within three to five years as humanoid robots achieve greater capability and cost competitiveness.[26] Workers in their twenties have time to acquire new credentials and shift career trajectories, while those in their fifties face less options and greater urgency. Yet even older workers possess institutional knowledge, relationship networks, and judgment refined through decades of experience—assets that retain value if deliberately repositioned toward new emerging roles.

The humanoid revolution will not wait for perfect preparation, but the difference between thriving and merely surviving this transition depends enormously on choices made today. Workers who build adaptable skills, families who create financial resilience, and communities that invest in diversification and education, position themselves not as automation's victims but as participants in shaping what the robot age actually looks like and who it serves. The transformation of work done by humanoid robots will not unfold as a single dramatic event but as a series of shifts—some sudden, others gradual—that will reshape employment landscapes over the coming decade. The workers who stock warehouse shelves today, the manufacturing employees who assemble products on factory floors, and the retail associates who manage inventory face genuine disruption that no amount of optimistic rhetoric can dismiss. Yet the outcome of this transformation remains unwritten, dependent not on technological inevitability but on choices that

workers, employers, communities, and policymakers make right now about preparation, investment, and values.The future is shaping up to be more complex than either dystopian warnings or utopian promises suggest. Some jobs will indeed disappear with positions involving routine physical tasks in predictable environments facing the most immediate vulnerability, and workers lacking college degrees and those in already-struggling regions bearing disproportionate burdens. The research tracking robot deployment shows real job losses concentrated among populations least equipped to absorb them: men in manufacturing communities, non-White workers, and those without higher education credentials. These aren't statistics but neighbors, family members, and fellow community members whose livelihoods deserve more than dismissive assurances that "technology always creates new jobs eventually."

Yet alongside displacement, genuine opportunities for human-robot collaboration are emerging in warehouses, hospitals, manufacturing facilities, and logistics centers where early adopters are discovering that the most productive arrangements leverage what each partner does best. Humans freed from repetitive physical labor can focus on quality control, problem-solving, customer relationships, and the creative judgment that machines cannot replicate. The assembly worker becomes a process improvement specialist; the warehouse employee transitions to operations coordination; the healthcare aide focuses on emotional support while robots handle continuous monitoring. These hybrid roles aren't hypothetical futures but realities today in facilities already deploying humanoid robots and demonstrating that augmentation rather than replacement remains achievable when organizations commit genuinely to workforce development rather than simply cost reduction.

The practical strategies for navigating this shift require honest self-assessment, deliberate skill-building, and community-level investment that cannot wait for perfect information or ideal circumstances. Workers must evaluate which aspects of their roles face automation risk and cultivate complementary capabilities to enable them to work alongside robotic systems, and elevate their human skills like empathy, creativity, and complex judgment to remain beyond robotic reach. Families need income diversification and financial resilience to weather transitions that may require months of retraining. Communities dependent on vulnerable industries must invest now in educational infrastructure, economic diversification, and support systems that prevent the cascading decline previous automation waves triggered in unprepared regions.

The timeline for these preparations grows shorter as humanoid robots approach cost parity with human labor and deployment accelerates across multiple sectors. The window between "robots are coming eventually" and "robots are here now" is closing, particularly for workers in routine manual positions. Yet even compressed timelines allow for meaningful action when approached with urgency and purpose rather than paralysis or denial.

The fundamental question this chapter poses isn't whether humanoid robots will change work but whether that change fosters human flourishing or merely corporate efficiency, whether disruption's costs fall on those least able to bear them or get shared across society, and whether workers face this transformation as passive victims or active participants shaping outcomes. The robot revolution in the workplace is inevitable, but the world of work it creates remains ours to determine through choices we make today about education, policy, investment, and the values we refuse to compromise regardless of technological capability.

CHAPTER 4. HUMANOID ROBOTS IN EVERYDAY LIFE: TRANSFORMING HOMES, ROUTINES, AND DAILY LIVING

T he gap between what robots can do in controlled laboratory settings and what they can handle in the chaos of real homes with pets underfoot, children's toys scattered across floors, and the unpredictable rhythms of actual family life remains vast, and understanding that gap matters more than any glossy promotional video showing a robot folding laundry in a pristine showroom. This chapter explores the messy, complicated, deeply human reality of what it might actually mean to share our most intimate spaces with humanoid machines.

When discussions turn to humanoid robots in homes, the conversation often splits into two extremes: breathless excitement about robot butlers handling every household chore, or dystopian warnings about surveillance machines invading our privacy and replacing human connection. The truth, as usual, lives somewhere in the complicated middle in a place where genuine benefits coexist with real limitations, where convenience creates new dependencies, and where the same technology that helps one person maintain independence might make another feel diminished or monitored. Navigating this middle ground requires looking past both the hype and the fear to examine what humanoid robots can realistically offer in homes today and in the near future, what they cannot do despite ambitious promises, and what trade-offs we're actually making when we invite them across our thresholds.

The home represents fundamentally different territory than the workplace environments where robots have already proven their value. Factories and warehouses offer structured, predictable settings where robots perform repetitive tasks in controlled conditions, but homes are gloriously chaotic and filled with variables no engineer can fully anticipate. A humanoid robot might navigate a clear hallway beautifully but struggle when a toddler leaves a stuffed animal in its path, or when an elderly resident's walker sits at an unexpected angle, or when the family dog decides the robot's leg looks like an interesting thing to investigate. These aren't minor technical details—they're the difference between a robot that genuinely helps and one that creates more frustration than it solves.

Yet for all these limitations, the potential for humanoid robots to meaningfully support daily living—especially for aging adults, people with disabilities, and overwhelmed caregivers—deserves serious attention rather than dismissal. The question isn't whether robots will transform domestic life, but how that transformation will unfold, who will benefit most, what we'll gain, and what we might lose in the process. This chapter examines the practical realities of humanoid robots in homes through three critical lenses: what they can actually do to support household tasks and daily routines, how they might help older adults age in place with dignity and independence, and the privacy concerns that emerge when cameras and sensors become permanent residents in our most private spaces. By grounding these discussions in real caregiving challenges, actual family dynamics, and honest assessments of current technology, readers can develop realistic expectations and make informed decisions about if, when, and how to integrate humanoid robots into the complicated, beautiful, messy reality of their home.

From Chores to Companionship: What Humanoid Robots Can (and Can't) Actually Do in Your Home

The promise of a robot that folds laundry, vacuums floors, and prepares simple meals has captivated imaginations for decades, but the reality arriving in homes today looks quite different from those science fiction fantasies. As of late 2025, the first consumer-ready humanoid robots have begun entering real households, marking a genuine turning point in domestic technology. Yet understanding what these machines can actually accomplish versus what marketing materials suggest remains essential for anyone considering inviting one across their threshold.[27][28]

1X's NEO represents the vanguard of this domestic robot revolution, marketed explicitly as the world's first consumer-ready humanoid robot designed for home life.[28] The company began testing NEO in hundreds to thousands of actual homes by the end of 2025, with real consumer preorders opening that October. NEO arrives with capabilities that sound modest but prove meaningful in daily life: opening doors for guests, fetching items from other rooms, controlling smart home devices like lights and thermostats, and holding basic conversations.[28] What distinguishes NEO from earlier robotic attempts is its foundation in artificial intelligence that allows it to learn and expand its skills through continued use and software updates, rather than remaining locked in its factory programming.[28]

Other robots entering the domestic market include **Neo Gamma** from Norwegian company 1X Technologies, which demonstrated during showcases at NVIDIA GTC 2025 household tasks including vacuuming, watering plants, and navigating living spaces without colliding with furniture or people. **Clone Robotics** planned to release 279 units of its Clone Alpha in 2025 at approximately twenty thousand dollars, with promised capabilities including laundry, dishwashing, and cooking basic meals. These early models share a common characteristic: they excel at specific, well-defined tasks but struggle with the contextual reasoning and adaptability that humans take for granted.

The gap between demonstration videos and daily reality matters enormously. Recent showcases by Google DeepMind featured robots handling clothes, sorting items into bins, and responding to natural language commands—all genuine technological progress that researchers describe as "a step in the right direction."[29] Yet these demonstrations occur in controlled environments with carefully selected tasks, and the leap from performing specific actions in optimized scenarios to autonomous, general-purpose domestic assistance remains substantial.

Current humanoid robots face critical limitations that constrain their usefulness in real homes. Unlike industrial robots operating in predictable factory settings, domestic robots must navigate gloriously chaotic environments filled with pets, children's toys scattered across floors, unexpected obstacles, and constantly changing layouts. A robot might successfully vacuum a clear hallway but struggle when a chair sits at an unexpected angle or when a cat sits in front of it and refuses to move.

The ability to simulate human senses represents perhaps the most significant humanoid bottleneck, particularly regarding tactile feedback and touch sensing.[29] While vision improves with extensive training datasets gathered from internet sources, comparatively little data exists for the sense of touch which is essential for manipulating both soft and hard objects safely.[29] Current

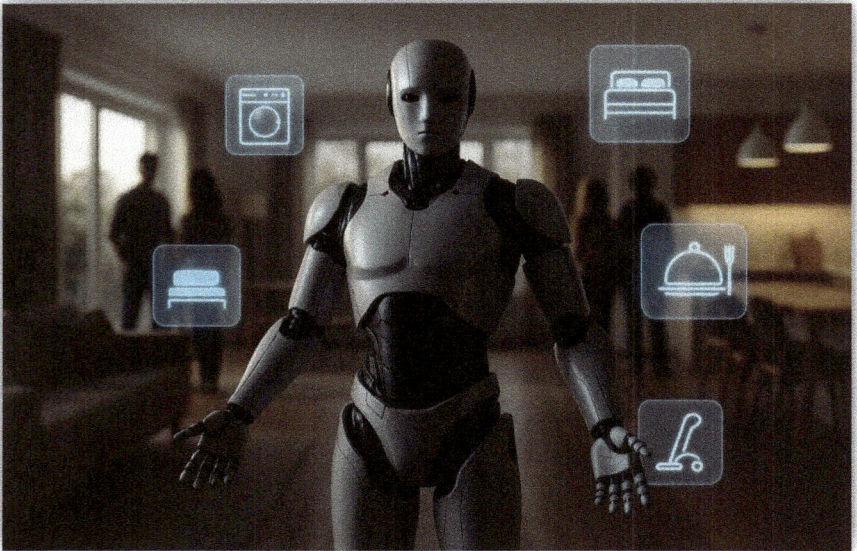

day robots also cannot register pain or smell, which are critical senses that humans rely upon to make decisions in uncertain environments.[29] Researchers are developing electronic robot skins to provide tactile feedback, but these technologies remain in development rather than deployment.[29]

For families considering whether a humanoid robot might genuinely help with daily life, realistic expectations matter more than promotional hype. Current robots work best for specific, repeatable tasks: turning off lights at bedtime, fetching items from designated locations, basic tidying in uncluttered spaces, and serving as hubs for smart home device control. They struggle with tasks requiring fine motor control, contextual judgment about when assistance is needed, being able to adapt to unexpected situations, and anything involving complex social dynamics or nuanced decision-making. The robot that can reliably fold a fitted sheet, prepare a meal from whatever ingredients happen to be available, or recognize when someone needs help versus honoring their privacy remains years away from reality, despite what glossy marketing materials might suggest.

Aging in Place with Robotic Support: How Humanoids Could Help Seniors Maintain Independence and Dignity

The desire to remain in our own home as the body ages and our abilities decline represents one of the most universal human wants yet achieving that goal grows increasingly difficult as populations age faster than the caregiving workforce can expand. By 2030, the number of adults over eighty will reach unprecedented levels globally, while simultaneously, the availability of professional in-home caregivers continues to contract, creating a care crisis that leaves families scrambling for solutions and seniors facing unwanted moves to institutional settings.[32] Humanoid robots, for all their limitations in other domestic applications, may offer their most compelling value proposition precisely here: supporting older adults who want to age in place with dignity, safety, and genuine independence rather than surveillance disguised as care.[30] [32]

The practical challenges that push seniors out of their homes often involve tasks that humanoid robots can realistically address with current or near-future technology. Medication management—remembering which pills to take when, tracking refills, and maintaining consistent schedules—becomes increasingly difficult for humans as cognitive function declines, yet represents exactly the kind of structured, repeatable task that robots handle well.[30] A humanoid robot can provide medication reminders at prescribed times, visually confirm that medications have been taken, and alert family members or healthcare providers when doses are missed, all without the cost or scheduling constraints of human home health aides.[30] Similarly, household tasks that become dangerous or impossible with declining mobility—reaching high shelves, carrying heavy items, navigating stairs with laundry—fall within the emerging capabilities of robots like 1X's NEO or Clone Alpha, which can fetch items, transport objects between rooms, and handle basic cleaning tasks in ways that preserve a senior's ability to maintain their living space without constant outside help.[30]

Fall prevention represents perhaps the most critical application where robotic support directly enables aging in place. Falls currently stand as the leading cause of injury in adults over sixty-five, and the fear of falling, sometimes even more than falls themselves, often drives the decision to leave home for a supervised care setting.[31] MIT's development of the E-BAR (Eldercare robot for Balance Assistance and Rehabilitation) addresses this challenge by providing robotic handlebars that stabilize the body during standing, sitting, and movement, following users throughout their homes and offering support whenever needed.[31] The design acknowledges a psychological reality that traditional walkers and canes cannot overcome: many older adults underestimate fall risk and refuse cumbersome physical aids, while others overestimate risk and avoid movement entirely, leading to declining mobility that becomes self-fulfilling prophecy.[31] A robot that provides unobtrusive support without the stigma of traditional assistive devices may bridge this gap, allowing seniors to move confidently through their homes while genuinely reducing fall risk.[31]

Yet the most profound contribution humanoid robots might make to aging in place is to address the isolation and loneliness that often accompany growing old, particularly for those living alone. Companion robots equipped with conversational AI can engage seniors in daily interaction—not replacing human connection, but filling the long hours between family visits or caregiver shifts when silence and solitude weigh heavily.[33] Research consistently shows that older adults view robots as tools for assistance rather than replacements for human relationships, a distinction that matters enormously for acceptance and effectiveness.[30] Seniors recognize the difference between a robot's programmed responses and a grandchild's genuine affection, but they also recognize that a robot offering cognitive games, reminiscence conversations, or simple presence provides something valuable in moments when human company isn't available.[30]

The ENRICHME project demonstrates how targeted robotic intervention can support specific populations within the aging community, focusing on older adults with mild cognitive impairment who face particular challenges maintaining independence.[30] By providing cognitive training, exercise facilitation, and environmental monitoring tailored to their needs, these robots extend the window during which individuals can safely remain at home— buying precious time for families navigating difficult care decisions and preserving quality of life during a vulnerable transition period.[30] This represents the realistic promise of aging-in-place robotics: not miracle solutions that eliminate all challenges of growing old, but practical tools that extend independence, reduce caregiver burden, and honor the deep human desire to remain in familiar surroundings surrounded by a lifetime's accumulation of memories, routines, and meaning.[30 32]

The Privacy Paradox: Cameras, Data, and Trust When Robots Live Where You Live

The moment a humanoid robot crosses the threshold into a home, it brings with it something most families don't immediately consider: a constellation of cameras, microphones, and sensors designed to help the robot navigate and respond, but also capable of capturing every intimate detail of domestic life. This creates what researchers call the privacy paradox. People express genuine concern about surveillance and data collection, yet these worries often fail to prevent them from welcoming robots into their most private spaces when the promise of convenience or assistance feels compelling enough.

Research on social robots reveals this contradiction clearly: individuals voice significant privacy concerns in surveys, yet these concerns disappear when actual robots become available.[35] Studies using realistic robot scenarios demonstrate that privacy-invasiveness does negatively affect people's intentions to use robots, but this concern is frequently dismissed when faced with the practical benefits a robot offers—particularly for families caring for aging relatives or managing overwhelming household demands.[35] The gap between stated privacy values and actual behavior matters enormously, because it suggests that privacy protections cannot rely solely on consumer choice or market forces to develop appropriately.

The privacy risks embedded in household robots extend across multiple dimensions that most consumers don't fully understand when making purchase

decisions. Modern humanoid robots require connection to home networks to function effectively, enabling remote monitoring and cloud-based intelligence that makes them genuinely useful, but this same connectivity creates vulnerabilities where unauthorized network access can compromise the entire robot and everything it has collected.[34] Many users never change the default passwords like "admin" or "0000" that manufacturers provide, leaving systems exposed to relatively straightforward hacking attempts.[34] Once compromised, a robot becomes an agent of surveillance within the intimate space of the home, potentially collecting and transmitting passwords, financial information, health data, and behavioral patterns without the owner's knowledge or consent.[34]

Beyond network security, the sensor architecture itself presents profound privacy challenges. Household robots equipped with cameras, microphones, 3D sensors, and increasingly sophisticated biometric monitors collect vast amounts of sensitive personal data: lifestyle patterns, floor plans, details of personal belongings, geographical coordinates within the home, online credentials, and health status information.[34] Unlike stationary security cameras that monitor specific areas, humanoid robots move freely through homes, entering bathrooms, bedrooms, and other traditionally private spaces where their presence fundamentally alters the nature of privacy.[35] Research found that informational privacy concerns dominated discussions about robots in the home, with participants focusing on what data robots collect, who has access to that information, and how it might be used—questions that often lack clear answers from manufacturers.[35]

The emotional bonds humans form with robots create an additional vulnerability that extends beyond technical security concerns. Scientific research demonstrates that people develop genuine attachments to household robots—particularly those designed for caregiving or companionship—and this emotional connection can be exploited.[34] A compromised robot that a family has come to trust can generate distressing sounds, move objects in threatening ways, or engage in strange behaviors that cause worry and concern.[34] For elderly individuals or children who rely on robotic assistance, this represents not just a privacy breach but a betrayal of trust in a relationship they've come to depend upon for daily support and safety.[34]

Perhaps most troubling, privacy vulnerabilities can originate not from external hackers but from robot vendors themselves. Service robots that continuously surveil homes as part of their functions have access to photographs, videos, location information, and behavioral data, and users must trust that manufacturers handle this access ethically, store data securely, and don't exploit their privileged position to collect information beyond what's necessary for robot function.[36] This represents a fundamental trust problem: families must rely on vendor ethics and security practices to protect their most intimate information, yet they have limited visibility into how manufacturers actually handle data access, storage, and sharing.[36]

The path forward requires moving beyond the privacy paradox by embedding protection into robot architecture from inception rather than treating it as an afterthought. Users need transparent documentation of exactly what sensors their robot contains and what data each one collects, along with granular controls to disable sensors not required for specific functions, notification systems that alert them when data is accessed, and clear governance explaining how information is stored and who can access it.[34] [37] Until these protections become standard rather than optional, families welcoming humanoid robots into their homes are making a privacy trade-off they may not fully understand—exchanging intimate access to their daily lives for the convenience and assistance these machines promise to provide.The promise of humanoid robots transforming domestic life sits at a peculiar crossroads—close enough that families can now preorder actual consumer models, yet far enough that the gap between marketing promises and daily reality remains vast and consequential. Understanding this distance matters more than any promotional video showing pristine robots folding laundry in showrooms that bear no resemblance to the chaotic, beautiful mess of real homes where the unpredictable rhythms and obstacles of actual family life unfold moment by moment.

What emerges from examining current capabilities is a picture neither as revolutionary as enthusiasts claim nor as disappointing as skeptics suggest. Humanoid robots entering homes in 2025 and beyond can genuinely help with specific, structured tasks—medication reminders that prevent dangerous missed doses, fetching items that spare aging joints from painful reaching, basic cleaning in uncluttered spaces, and smart home control for those humans who find the interfaces confusing or inaccessible. These contributions, while modest compared to science fiction fantasies of robot butlers handling every household need, prove meaningful precisely for the populations who need them most: older adults fighting to maintain independence in their own homes, people with disabilities navigating spaces designed without their limitations in mind, and overwhelmed caregivers desperately seeking any support that might ease impossible burdens.

The aging-in-place applications represent where current technology aligns most powerfully with genuine human need. The crisis facing families as populations age faster than caregiving capacity can expand creates urgent an demand for solutions that preserve dignity, independence, and the profound human desire to remain surrounded by a lifetime's accumulation of memories and meaning. Robots that prevent falls through unobtrusive support, provide cognitive engagement during long hours of solitude, and handle physical tasks that become dangerous for humans with declining mobility, offer something valuable. It's not a replacement for human connection, but practical tools that extend the window during which individuals can safely remain in familiar surroundings. This matters enormously for quality of life during vulnerable transitions, buying precious time for families navigating difficult care decisions while honoring the deep wish to age at home rather than in institutional settings.

Yet every benefit these machines offer arrives bundled with privacy trade-offs that most families don't fully understand when welcoming robots across their thresholds. The cameras, microphones, and sensors that enable robots to navigate and respond also capture intimate details of domestic life—creating vulnerabilities when security is compromised and transforms helpful assistants into surveillance agents. The privacy paradox—where people express genuine concern about the security risk yet still adopt robots when convenience feels compelling enough—suggests that protection cannot rely on consumer choice alone. Until transparent data practices, granular sensor controls, and robust security become standard rather than optional features, families are making privacy exchanges they may not recognize until the consequences arrive uninvited.

The path forward requires moving beyond both breathless enthusiasm and reflexive fear toward clear-eyed assessment of what humanoid robots can realistically contribute to domestic life, what limitations constrain their usefulness, and what safeguards must exist before widespread adoption. For readers considering whether these machines might genuinely help their households, the question isn't whether robots will transform daily living (they already are) but whether that transformation will serve human flourishing or simply add technological complexity to lives already overwhelmed by devices demanding attention, data, and trust. The answer depends less on what engineers build and more on what families, communities, and societies demand: robots designed to augment human capacity and preserve dignity rather than replace connection or compromise the intimate sanctuary that home represents in an increasingly surveilled world.

CHAPTER 5. ROBOTS AND ELDER CARE IN THE FUTURE: DIGNITY, INDEPENDENCE, AND COMPASSIONATE TECHNOLOGY

Mrs. Chen was eighty-seven years old when I first met her. Despite the limitations imposed by arthritis and balance issues that increased her risk of a fall in the kitchen and bathroom, she remained highly self-reliant. Her primary concern was not mortality or discomfort but rather institutionalization and the resulting loss of her established routines, her garden, and the autonomy to make daily choices such as when to rise and what to eat. She expressed a strong preference for accepting the risks associated with living alone over relinquishing the independence she valued in the home where she had raised four children. One afternoon, her daughter reached out to me, expressing distress at the dilemma she faced: balancing respect for her mother's wishes with anxiety about her safety with only being able to do twice-daily wellness checks. This situation illustrates the complex challenges many families encounter when balancing elder well-being, autonomy, with familial responsibility.

This tension sits at the heart of one of the most profound challenges facing modern society: how to care for a rapidly aging population when there simply aren't enough human caregivers to meet the need, when families are scattered across states or countries, when the cost of professional care exceeds what most people can afford, and when the people who need help most desperately want to maintain their independence and dignity. The numbers tell a stark story. By 2030, all Baby Boomers will be older than sixty-five, creating an elder care crisis of unprecedented scale. Rural communities face even steeper challenges, with younger generations moving away to find work and healthcare providers becoming increasingly scarce. The traditional family structures that once provided elder care have shifted, leaving millions of older adults navigating their aging with limited support systems.

Into this crisis, humanoid robots are emerging as a potential solution, and one that promises to help bridge the gap between the amount of care people need and the lack of human caregiver availability to provide it. These machines could help in a variety of ways such as monitoring for falls, reminding patients to take medications, assisting with mobility, providing companionship during long lonely hours, and alerting family members or medical professionals when something goes wrong. The technology is advancing rapidly, with robots already demonstrating capabilities that seemed impossible just a decade ago. For families like Mrs. Chen's daughter, the possibility of a robotic assistant that could check on her mother, help her move safely through her home, and provide an extra layer of security sounds like an answer to desperate prayers.

Yet this promise comes wrapped in profound questions that go far beyond engineering and into the deepest territory of what it means to care for another human being. Can a robot provide genuine companionship, or does it simply create the illusion of connection while deepening isolation? Does robotic assistance preserve dignity and independence, or does it become a cheaper, more convenient substitute for the human presence that aging adults actually need? Who decides when an older person should accept robotic help? The elder themselves? Worried family members? Or healthcare systems looking to reduce costs? And perhaps most troubling is the question of as we develop robots capable of performing caregiving tasks, do we risk creating a two-tiered system where wealthy families can afford human caregivers while everyone else gets machines?

This chapter explores these questions with the honesty they deserve, examining both the tremendous potential of humanoid robots in elder care and the ethical minefields we must navigate carefully. The goal isn't to determine whether robots belong in elder care (they're coming regardless) but rather to understand how we can deploy them in ways that truly serve aging adults, support rather than replace human caregivers, and ensure that technology enhances elder dignity rather than diminishing it.

The Elder Care Crisis: Why We Desperately Need New Solutions (and Why Robots Aren't the Only Answer)

The numbers tell a story that should alarm anyone paying attention. Sixty-three million Americans now serve as family caregivers—a staggering 45 percent increase over just the past decade.[38] This means roughly one in four American adults currently provides care for an aging or disabled loved one[38], often while simultaneously managing their own careers, raising children, and trying to maintain some semblance of financial stability. The value of this unpaid labor exceeds $870 billion annually[42], representing an invisible subsidy that keeps the entire healthcare system from collapsing under its own weight.

Behind these statistics lie millions of families facing impossible choices. Forty-five percent of older adult households (more than 19 million) lack sufficient income to cover basic living costs.[39] Sixty percent of older adults cannot afford two years of in-home care services[39], despite expressing strong preferences to age in their own homes rather than enter institutional settings. The wealth gap among seniors continues to widen, with devastating health consequences: older adults with the fewest financial resources die, on average, nine years earlier than their wealthier peers.[39]

The demographic pressure intensifying this crisis shows no signs of abating. In some regions, the population of people aged 85 and older—those most likely to require intensive care—is projected to increase by more than 70 percent over the next decade.[40] Approximately 20 percent of these "super seniors" will require assistance with the basic activities of daily living: bathing, dressing, eating, using the bathroom, managing medications.[40] Many will need round-the-clock monitoring to prevent falls, respond to medical emergencies, or simply provide the human presence that makes life bearable during long, isolated days.

The traditional safety net designed to catch vulnerable elders is fraying badly. The United States faces a critical shortage of direct care workers like home health aides, personal care assistants, and nursing assistants, who provide the hands-on support that enables aging in place.[41] Even states that have made targeted investments in workforce development struggle to maintain adequate staffing levels.[41] Meanwhile, fiscal pressures at both state and federal levels threaten to reduce funding for programs like Medicaid, which provides the essential support for long-term care services that most families cannot afford out of pocket.[41]

Perhaps most troubling is the emergence of the "sandwich generation"— individuals simultaneously caring for aging parents and dependent children. Twenty-nine percent of all caregivers fall into this category, and among caregivers under age 50, that proportion rises to 47 percent.[38] These caregivers report higher stress levels, reduced work hours, and diminished ability to save for their own retirements[43], which creates a cascading effect that threatens to

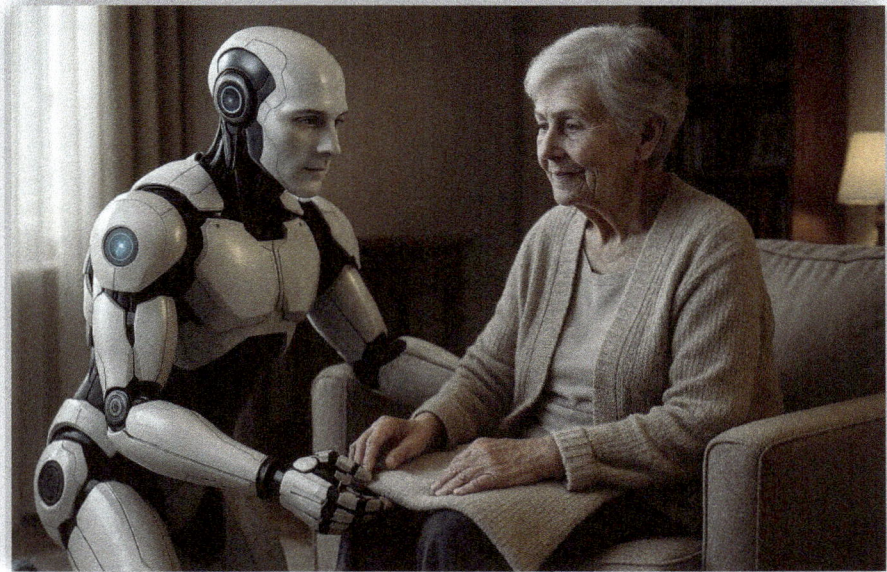

perpetuate economic insecurity across generations.

This crisis demands urgent, multifaceted solutions. Policy reforms must increase funding for long-term care services and provide meaningful support for family caregivers. Workforce development initiatives need sustained investment to recruit, train, and retain direct care workers while ensuring they receive fair compensation for their demanding, essential work. The goal should be to make care more affordable and accessible across income levels, so it's not just for the wealthy.

Technology represents one potential tool in this broader toolkit, but only one. Robots could assist with medication reminders, fall detection, mobility support, and companionship during lonely hours. They could also free human caregivers to focus on the emotional and relational dimensions of care that machines cannot replicate. Yet technology alone cannot solve a crisis rooted in policy failures, workforce shortages, and systemic financial constraints. The danger lies in treating robots as a convenient substitute for the human investment—both financial and relational—that genuine elder care requires. The question facing society is not whether robots can help, but whether their deployment will serve as a supplement to comprehensive care reform or as an excuse to avoid making the difficult policy choices and financial commitments that true solutions demand.

What Humanoid Robots Can Actually Do for Aging Adults: From Safety Monitoring to Cognitive Engagement and Companionship

Understanding what humanoid robots can actually accomplish for aging adults requires moving beyond both the hype of marketing materials and the skepticism of those who dismiss robots as cold, impersonal machines. The reality sits somewhere in between. These types of technologies offer capabilities that address real needs, but they work best when thoughtfully integrated into a broader ecosystem of human care rather than positioned as replacements for it.

Safety monitoring represents one of the most immediately practical applications of humanoid robots in elder care. Falls constitute the leading cause of injury-related deaths among adults over sixty-five, with one in four older Americans experiencing a fall each year. Humanoid robots equipped with advanced sensors can detect falls in real-time and immediately alert emergency contacts or medical services, dramatically reducing the dangerous window between incident and intervention. Beyond fall detection, these robots can monitor movement patterns throughout the home, identifying changes that might signal declining mobility or emerging health issues before they become crises.[44] For families separated by distance like a daughter in Seattle worrying about her father in rural Montana, this continuous monitoring provides reassurance without requiring constant phone calls or intrusive camera surveillance that many seniors find demeaning.

Health management capabilities extend beyond emergency response to include the daily routines that keep chronic conditions under control. Humanoid robots can provide medication reminders at scheduled times, verify that medications have been taken, and alert caregivers when doses are missed.[30] For seniors managing multiple prescriptions, which is a reality for the majority of Americans over sixty-five, this seemingly simple function prevents dangerous medication errors and hospitalizations. Some advanced systems can monitor vital signs including heart rate, blood pressure, and oxygen saturation[30][44] and transmit the data to healthcare providers and enabling proactive intervention when readings fall outside normal ranges. This continuous health surveillance transforms care from reactive (responding after problems emerge) to preventive (catching warning signs early when interventions are simpler and more effective).[44]

Cognitive engagement represents another domain where humanoid robots demonstrate measurable benefits, particularly for individuals experiencing mild cognitive impairment or early-stage dementia. Robots like Paro, the therapeutic seal used extensively in Japanese care facilities, have been documented providing reductions in stress and anxiety among dementia patients.[44] Humanoid robots can facilitate structured cognitive activities such as memory games, conversation exercises, reminiscence therapy, that help maintain mental acuity.[30] They can adapt difficulty levels based on individual performance, thereby providing personalized cognitive training that would be prohibitively expensive if delivered one-on-one by human therapists. For seniors living alone, these interactions provide mental stimulation during long hours when family members are at work or managing their own responsibilities.

Perhaps most controversial yet potentially most valuable is the companionship dimension of humanoid robots in elder care. Loneliness and social isolation affect millions of older adults, particularly those in rural areas or those who have outlived spouses and friends. Studies have documented that chronic loneliness carries health risks comparable to smoking fifteen cigarettes daily, increasing mortality risk and accelerating cognitive decline. Humanoid robots designed for social interaction can engage in conversation, recognize emotional states, and provide a responsive presence during lonely hours.[30] In nursing homes, socially assistive robots have demonstrated the ability to enhance social engagement and improve mood among residents.[30][45][46]

The critical question isn't whether robots can provide these functions, because evidence confirms they can, but rather what role they should play relative to human caregivers, and whether robotic assistance genuinely serves the dignity and preferences of aging adults themselves or merely the convenience and cost concerns of families and healthcare systems.

The Ethics of Care: Ensuring Robot Assistance Preserves Dignity, Choice, and Human Connection Rather Than Replacing It

The promise of humanoid robots in elder care rests on a delicate ethical foundation that can either support human flourishing or crumble under the weight of convenience and cost-cutting. When a robot reminds an older adult to take medication, monitors for falls, or provides companionship during lonely afternoons, these functions appear straightforward, as helpful actions that address real needs.[47] Yet beneath these practical applications lies a profound question: does robotic assistance preserve the dignity, autonomy, and human connection that make life meaningful, or does it gradually replace these irreplaceable elements with mechanical efficiency?

The answer depends entirely on how these technologies are deployed and who makes the decisions about their use. Research from Nordic caring traditions— ethical frameworks developed by scholars including Eriksson, Martinsen, and Dahlberg—emphasizes that genuine care must respect the patient's world in all its dimensions: the freedom to make decisions, express oneself, maintain privacy, engage in open communication, and safeguard psychological well-being alongside physical health.[47] These principles provide essential guardrails for evaluating whether robotic interventions enhance or undermine compassionate care.

Autonomy represents the first ethical cornerstone that robotic care must protect rather than erode.[47] When robots are introduced into care environments, decisions about their use, function, and interaction patterns are typically made by facility administrators, family members, or healthcare

professionals and not by the older adults themselves.[47] An elderly person may struggle to understand how to operate the technology, and this lack of technical literacy translates directly into a loss of control over their immediate environment and care experience. The robot becomes a tool through which third parties exercise control, responding according to programming determined by others rather than preferences expressed by the person receiving care.[47]

Preserving autonomy in robotic care settings requires more than simply placing a robot in someone's home. It demands genuine understanding. Older individuals need clear, accessible education about how these systems work and what they can and cannot do.[47] It requires personalization, allowing individuals to customize robotic behavior to match their preferences, routines, and values.[47] Without this foundation, robots risk becoming instruments of paternalism, making decisions for older adults rather than empowering them to maintain control over their own lives.

Privacy and dignity concerns become especially acute when robots perform intimate care tasks.[47] Unlike human caregivers who exercise discretion, sensitivity, and protective instincts regarding vulnerability, robots operate according to fixed protocols without understanding the deeper significance of privacy invasion. When a robot assists with bathing, dressing, or toileting—tasks traditionally performed by trusted human caregivers—the experience can feel objectifying rather than supportive. Older adults may feel reduced to bodies requiring maintenance rather than a whole person deserving respect, particularly when they're aware that someone else controls the robot's functions and potentially views the data it collects.[47]

The companionship paradox presents perhaps the most troubling ethical dilemma. Research confirms that human social networks and genuine companionship play crucial roles in delaying dementia onset and maintaining cognitive health in older adults.[47] Yet robots positioned as companions create a dangerous risk because while they provide the appearance of social interaction, they may actually reduce genuine human contact.[47] For older adults with mobility limitations or cognitive impairment (those least able to seek out alternative social connections) this substitution can lead to profound isolation masked by the illusion of companionship.

The fundamental problem isn't that robots engage in conversation or provide responsive presence during lonely hours. The problem emerges when robotic interaction displaces rather than supplements human connection, when families or facilities reduce human contact because "the robot keeps them company," or when older adults communicate less with human caregivers because the robot fulfills surface-level social needs.[47] What appears to be dialogue between robot and elderly person is actually a third party controlling the machine, determining its functions, providing its responses. It is a relationship fundamentally lacking the authentic reciprocity that genuine human connection requires.[47]

Attentive care characterized by empathy, compassion, and sensitivity to individual needs represents another dimension robots fundamentally

cannot replicate.[47] Human caregivers adapt their approach based on subtle emotional cues, past experiences with a patient, and intuitive understanding of complex needs. They recognize when someone needs comfort beyond physical assistance, when emotional support takes precedence over task completion, or when the most caring response is simply sitting with someone in their distress. By contrast, robots tend to react in standardized ways regardless of individual differences, as they are lacking the cognitive and emotional capacities required for genuine attentiveness to suffering, fear, confusion, or grief.[47]

The ethical path forward requires positioning robots as complements to human caregivers rather than replacements—machines that handle routine, physically demanding tasks while the human caregivers focus on providing emotionally responsive, relationally rich care.[47] This model only works, however, if institutions genuinely invest in maintaining human caregiving capacity rather than using robots as justification for reducing staff. The choice facing families, healthcare facilities, and policymakers isn't whether robots can assist with elder care, but whether their deployment will serve the dignity and wellbeing of older adults or merely serve institutional convenience and cost reduction.The elder care crisis facing modern society demands urgent, comprehensive solutions, such as more direct care workers, better compensation for caregiving labor, increased funding for long-term care services, and meaningful support for the millions of family caregivers shouldering impossible burdens. Humanoid robots represent one potential tool in addressing this crisis, capable of monitoring safety, managing health routines, providing cognitive engagement, and offering companionship during lonely hours. Yet their value depends entirely on whether they're deployed as supplements to human care or substitutes for it, and whether they serve the dignity and preferences of aging adults or merely the convenience and cost concerns of institutions and families.

The capabilities these robots offer are genuine and measurable. Fall detection systems can dramatically reduce the dangerous window between incident and intervention, potentially saving lives and preventing serious injuries. Medication management functions help seniors maintain independence while avoiding dangerous errors that lead to hospitalizations. Cognitive engagement activities provide mental stimulation that helps preserve function and delay decline. For families separated by distance or overwhelmed by competing demands, these technologies offer reassurance and practical support that can make the difference between sustainable caregiving and complete burnout.

Yet every benefit comes wrapped in ethical complexity that cannot be ignored or dismissed. Autonomy requires that older adults themselves (not just their families or care facilities) make informed decisions about robotic assistance by understanding how these systems work and maintaining control over their

use. Privacy and dignity demand careful attention to the intimate nature of caregiving tasks and the vulnerability inherent in accepting help with basic activities of daily living. The companionship paradox presents perhaps the most troubling challenge: while robots can provide responsive presence and conversation, they risk displacing genuine human connection rather than supplementing it, creating isolation masked by the illusion of social interaction.

The path forward requires clear-eyed commitment to principles that place human flourishing above technological convenience. Robots should handle routine monitoring and physically demanding tasks, freeing human caregivers to focus on emotionally responsive, relationally rich care that machines cannot replicate. Older adults must retain meaningful control over how and when robots assist them, with education and personalization ensuring technology serves their preferences rather than overriding them. Investment in robotic assistance must never become a justification for reducing human caregiving capacity in families, facilities, or healthcare systems, but instead should enable more sustainable, comprehensive care that honors both practical needs and irreplaceable human dimensions.

Mrs. Chen's daughter faced an impossible choice between her mother's fierce independence and legitimate safety concerns, between honoring autonomy and preventing tragedy. A humanoid robot monitoring for falls, assisting with mobility, and providing an extra layer of security might have eased that impossible tension, but only if it preserved Mrs. Chen's control over her own home, protected her privacy and dignity, and supplemented rather than replaced her daughter's visits and genuine human connection. The technology exists to help families navigate these painful dilemmas. Whether it actually serves aging adults or merely serves institutional convenience depends on choices society makes now, before these machines become ubiquitous, about what values will guide their deployment and whose needs they will truly prioritize.

CHAPTER 6. CHILDREN AND HUMANOID ROBOTS: GROWING UP IN A ROBOT-INTEGRATED WORLD

My neighbor's five-year-old daughter has never known a world without voice assistants, and she says "please" and "thank you" to Alexa with the same careful politeness her parents taught her to use with human adults. Watching her navigate this hybrid social world where some companions are flesh and some are code and where kindness is practiced on machines and humans alike, I realize we're raising the first generation who won't remember a time before robots were part of childhood, and we're making up the parenting rules as we go. What happens when the voice assistant gets a body? What about when the helpful technology takes human form and walks into classrooms, playrooms, and bedrooms where children are learning what relationships look like and how the world works?

This question keeps parents, educators, and pediatric specialists awake at night, and for good reason. Children's brains are exquisitely designed to learn from their environment, to form attachments, to practice social skills through thousands of daily interactions, and we're about to introduce a fundamentally new type of social entity into that developmental landscape without a roadmap, longitudinal studies, or cultural precedent to guide us. Unlike previous technological shifts that changed what children did (television, video games, smartphones), humanoid robots will change who children interact with and potentially alter the very foundation of how they learn to be human among other humans.

The stakes are genuinely high, but so is the potential. Humanoid robots could offer personalized educational support that adapts to each child's learning style and pace, provide patient companionship for children with social anxiety or autism spectrum disorders, and serve as tireless tutors in under-resourced schools where teacher shortages leave students behind. They could help children with disabilities gain independence, offer language practice for immigrant families, and create learning experiences that no textbook or screen can replicate. The same technology that worries us might also unlock opportunities we've barely imagined.

Yet we cannot let enthusiasm override caution. The research on children and screen time offers sobering lessons about unintended consequences and how technologies designed to educate often end up replacing active play, face-to-face conversation, and the beautiful inefficiency of human childhood. Precisely because they're more engaging and lifelike than screens, humanoid robots carry even greater risks of displacing the human interactions that wire developing brains for empathy, creativity, and genuine connection. A robot tutor that never gets frustrated might sound ideal until we realize that learning to navigate a teacher's bad mood is itself a crucial life skill.

This chapter examines the complex reality of childhood in a robot-integrated world with the honesty parents and educators deserve. We'll explore what developmental science tells us about how children form attachments and learn social skills, what early evidence suggests about kids interacting with humanoid robots, and where the genuine benefits lie versus where we're simply automating childhood in ways that serve adult convenience more than children's flourishing. We'll confront the equity issues that could make robot-enhanced education another privilege gap, and we'll provide practical frameworks for parents navigating decisions about when, how, and why to introduce humanoid robots into their children's lives , while always keeping the question front and center: does this technology serve the child's development, or does it simply make parenting easier while potentially compromising what childhood is meant to be?

The Developmental Question: How Robot Interaction Affects Social Skills, Empathy, and Human Relationship Formation in Children
Human development unfolds through thousands of small interactions like a parent's responsive smile, a sibling's teasing, a teacher's patient correction, or a friend's shared secret. These moments, repeated across years, wire the brain for empathy, teach the subtle rules of conversation, and build the capacity for

them to read emotions in others' faces and voices. When humanoid robots enter this developmental landscape, they introduce a fundamentally new category of social partner: entities that look human, respond to children's actions, and engage in back-and-forth exchanges, yet lack genuine feelings, consciousness, or authentic relationship capacity. The question facing parents and educators is not whether robots will affect child development, but how, and whether those effects serve children's long-term flourishing.

Research from multiple universities and clinical settings provides encouraging evidence that humanoid robots can support specific developmental outcomes, particularly for children facing social challenges.[52] Studies conducted with robots like NAO, ZENO, and KASPAR demonstrate measurable improvements in joint attention—the ability to share focus on an object or activity with another person—among children with autism spectrum disorder.[53] [51] [49] These children often struggle with the unpredictability and sensory complexity of human interaction, and find the structured, patient, and consistent behavior of robots less overwhelming.[50] The robot becomes a practice partner where mistakes carry no social penalty, where conversations can be paused and repeated without frustration, and where the child controls the pace of interaction in ways rarely possible with human peers.[50]

The developmental benefits extend beyond autism interventions. Humanoid robots equipped with language learning capabilities have demonstrated effectiveness in helping children practice pronunciation, engage in conversational role-play, and develop vocabulary in both native and foreign languages.[48] The robot's infinite patience such as its willingness to repeat the same word fifty times without irritation, creates learning conditions that even the most dedicated human teacher cannot consistently maintain. For children with speech delays, social anxiety, or learning differences, this judgment-free practice environment can build confidence that transfers to human interactions.[48]

Yet there are also important limitations and concerns that need to be considered, as well. Studies consistently show that while children with autism improve in specific skills like turn-taking and following instructions during robot interactions, gains in spontaneous imitation—the natural copying of others' behaviors that typically developing children do effortlessly—remain limited.[53] This suggests robots excel at teaching discrete, structured social

skills but may not fully replicate the rich, messy, improvisational learning that happens in genuine human relationships. A robot can teach a child to make eye contact on command, but it cannot teach the intuitive sense of when eye contact is comfortable versus when it feels intrusive, or how to read the subtle shift in a friend's expression that signals hurt feelings beneath polite words.

Developmental psychologists are primarily concerned about robots used for displacement rather than supplementation. A robot tutor that helps a child with autism practice greetings for twenty minutes daily represents valuable therapeutic support. A robot companion that becomes a child's primary social partner and replaces playground time and family conversation, represents a developmental risk.[52] Children learn empathy not through programmed responses but through experiencing genuine emotional reciprocity—seeing their actions affect another person's feelings, navigating repair after a conflict, experiencing the vulnerability of being truly known by another human being. No matter how sophisticated their programming, robots cannot provide this authentic emotional exchange.

The anthropomorphic design of humanoid robots—their human-like faces, expressive gestures, and responsive behaviors—creates both opportunity and risk.[49] Children naturally attribute mental states and feelings to human-like entities, which is a tendency that makes robots engaging and relatable but also potentially deceptive.[49] When a robot responds to a child's story with programmed sympathy, the child may experience this as genuine caring, forming an attachment to an entity incapable of reciprocating authentic concern. This asymmetry—the child investing real emotion in a relationship the robot cannot truly participate in—raises questions about whether such interactions teach healthy relationship patterns or subtly distort children's understanding of what genuine connection requires.

Humanoid Robots as Educational Tools: Personalized Learning, Special Needs Support, and the Risk of Deepening Inequality
The classroom robot doesn't get tired when a child asks the same question for the tenth time. It doesn't show frustration when a student struggles with a concept their peers mastered weeks ago, and it never runs out of patience for the child who needs extra time, different explanations, or a completely individualized approach to learning. This tireless adaptability represents the genuine potential of humanoid robots in education—not by replacing teachers,

but by providing the kind of supplementary personalized attention that human educators, no matter how dedicated, cannot physically sustain for thirty students simultaneously.

Multiple educational institutions have measured the benefits of humanoid robots entering classrooms as teaching assistants. Their studies show that AI-powered robots can adapt lesson difficulty, pacing, and content based on each student's responses and demonstrated comprehension levels, creating truly individualized learning pathways that traditional classroom instruction cannot replicate.[54]

The impact proves particularly significant for students with special needs. Research published in peer-reviewed journals indicates that up to twenty percent of children with autism spectrum disorder demonstrated significant behavioral and social improvements after interacting with humanoid robots.[55] These machines offer something uniquely valuable for neurodivergent learners: absolute predictability and consistency. Children with autism often find the unpredictability of human interaction like the subtle mood shifts, the varying energy levels, and the unspoken social expectations that change without warning can be overwhelming. Humanoid robots like NAO and PEPPER provide structured, patient interactions free from these variables, creating psychologically safe spaces where students can practice social skills, develop communication patterns, and build confidence without fear of judgment or social penalty.[48]

Schools implementing systems like EDUBOT in rural Indian communities report enhanced student participation and improved concept retention, particularly in regions facing chronic teacher shortages.[48] The robots handle repetitive instructional tasks—phonics drills, math practice, vocabulary building—and liberating human teachers to focus on creative problem-solving, emotional connection, and the complex mentorship that only humans can provide.[54] In linguistically diverse regions, multilingual robots bridge communication gaps, teaching in multiple languages and ensuring that students whose home languages differ from instructional languages don't fall behind.[54]

Yet this transformative potential carries a troubling shadow: the risk that humanoid robots will deepen rather than bridge educational inequality. The robotics technology requires significant capital investment, reliable electricity,

internet connectivity, and ongoing technical support, which are resources that affluent school districts possess and under-resourced schools lack. The regions most desperately needing educational support—rural areas with teacher shortages, underfunded urban schools, communities serving economically disadvantaged populations—are precisely those least likely to be able to afford these systems.[54]

Without intentional policy intervention and equitable funding models, humanoid robots could create a two-tiered educational system where wealthy students receive personalized, robot-enhanced instruction while disadvantaged students remain dependent on overcrowded classrooms and overstretched teachers. The very technology positioned as a solution to educational inequity risks becoming another marker of privilege and widening the opportunity gap it promised to close. Organizations like PredictML.ai are attempting to address this through partnerships with NGOs and explicit focus on rural deployment, but these efforts remain exceptions rather than norms.[54]

The fundamental question facing educators, policymakers, and communities is not whether humanoid robots can improve education (evidence suggests they can) but whether societies will ensure this improvement reaches all children or only those already advantaged by wealth and geography. The answer will determine whether these machines become instruments of educational equity or monuments to deepening inequality.

Parenting in the Robot Age: Setting Boundaries, Teaching Critical Thinking, and Ensuring Technology Serves Childhood Rather Than Replacing It

The parent's dilemma in the robot age centers on a deceptively simple question: how do we allow children to benefit from technological advances without surrendering the irreplaceable elements of human childhood? The answer requires neither wholesale rejection nor uncritical embrace, but rather intentional boundary-setting, deliberate cultivation of critical thinking, and constant vigilance about whether technology serves development or simply makes parenting more convenient.

Prior research on the impacts of children's screen time offers cautionary lessons that are directly applicable to humanoid robots. Studies consistently demonstrate that excessive digital media consumption correlates with reduced face-to-face interaction, decreased physical activity, and compromised sleep quality—all factors affecting cognitive and emotional development. Humanoid robots, precisely because they're more engaging and lifelike than screens, carry amplified versions of these risks. A tablet can be ignored; a humanoid robot that walks into a child's room, speaks their name, and responds to their emotions commands attention in ways that make disengagement psychologically harder.

Parents navigating this landscape need frameworks for decision-making that prioritize child development over technological novelty. The first question should never be "what can this robot do?" but rather "what developmental need does my child have, and is this robot the best way to address it?" A robot offering language practice for a child learning English as a second language serves a clear developmental purpose. A robot positioned as a child's "best friend" to reduce parental caregiving demands serves adult convenience while potentially compromising the child's capacity to form human attachments.

Establishing clear boundaries requires parents to distinguish between supplementation and replacement. Humanoid robots that supplement human interaction by providing additional practice, patient tutoring, or structured companionship while human relationships remain primary occupy fundamentally different territory than robots that replace human presence. A robot that helps a child practice reading for thirty minutes daily while parents prepare dinner supplements learning. A robot that becomes the primary responder to a child's emotional needs because parents are too busy or exhausted represents a replacement that compromises a child's healthy attachment formation.

The boundary-setting must extend to time limits, physical spaces, and emotional dependency. Just as thoughtful parents establish screen-free zones and times such as no devices at dinner and no screens in bedrooms, they should create robot-free spaces where human interaction remains protected. Bedtime routines, family meals, and unstructured outdoor play represent developmental contexts too valuable to outsource to machines, regardless of how sophisticated those machines become.

Teaching critical thinking about humanoid robots begins with transparency about what these machines actually are. Children benefit from understanding that robots, no matter how lifelike, are tools created by humans with specific purposes and limitations. Parents should demystify the technology through age-appropriate explanations: robots follow instructions written by programmers, their "feelings" are simulated responses rather than genuine emotions, and their patience stems from an inability to experience frustration rather than a natural virtue. This transparency doesn't diminish the robots' usefulness but prevents children from developing distorted expectations about relationships.

Critical thinking extends to questioning the robot's design and purpose. Parents can ask children to consider: Who created this robot? Why did they design it this way? What are they trying to accomplish? Does this robot help us do things we value, or does it replace things that matter? These questions cultivate the analytical capacity children will need as adults to navigate increasingly sophisticated artificial intelligence.[56]

The ultimate measure of whether technology serves childhood is whether it expands or contracts a child's capacity for human flourishing. Robots that free children to engage more deeply with creative play, physical activity, and human relationships serve development. Robots that become substitutes for the messy, demanding, irreplaceable work of human connection compromise it. Parents who maintain this distinction—prioritizing authentic human presence while selectively using robots for genuine developmental support—position their children to benefit from technological advances without sacrificing the foundations of healthy human development.

The children growing up today will never remember a world where machines couldn't talk back, where technology stayed safely contained in screens, or where the boundary between human and artificial companions was clear and uncrossable. This generation will navigate social landscapes populated by both flesh-and-blood friends and humanoid robots designed to teach, comfort, and engage with remarkable sophistication. The question facing parents, educators, and communities is not whether this reality will arrive, as it already has in early forms, but whether adults will guide this integration with the wisdom and intentionality that children's development demands.

There are genuine reasons for both hope and caution. Humanoid robots demonstrate measurable benefits for children with autism spectrum disorders, providing patient, predictable practice partners that reduce social anxiety and build communication skills that transfer to human interactions. Educational robots adapt to individual learning styles with a precision no human teacher managing thirty students can match, offering personalized support that could

make a huge difference for struggling learners. For children with disabilities, language barriers, or social challenges, these machines represent tools that expand rather than limit developmental possibilities.

Yet the same characteristics that make humanoid robots effective—their tireless patience, their engaging human-like presence, their ability to respond without judgment—also create risks that responsible adults cannot ignore. Children learn empathy through genuine emotional reciprocity, develop social skills through navigating the beautiful messiness of human relationships, and build resilience by experiencing the full range of human responses including frustration, disappointment, and repair. Robots that become primary social partners rather than supplemental tools risk displacing the very interactions that wire developing brains for authentic human connection.

Furthermore, without intentional policy intervention and equitable funding structures, humanoid robots could create educational divides where privileged children receive personalized, robot-enhanced instruction while disadvantaged students remain dependent on overcrowded classrooms and overstretched teachers. The technology positioned as a solution to educational inequality risks becoming another marker of privilege, widening opportunity gaps in communities already struggling with resource scarcity. Rural schools facing teacher shortages and underfunded urban districts serving economically disadvantaged populations are the ones that need these tools most yet can afford them least.

Parents navigating this landscape need frameworks that prioritize child development over technological novelty or adult convenience. The guiding principle should be supplementation rather than replacement—using robots to extend human capacity for teaching and support while protecting the importance of human relationships. This requires establishing clear boundaries around time, space, and emotional dependency, teaching children critical thinking about what robots actually are, and constantly asking whether technology serves the child's flourishing or simply makes parenting easier.

The stakes extend beyond individual families to the fundamental question of what childhood should be in an age of increasingly sophisticated artificial companions. Societies must decide collectively whether to protect unstructured play, face-to-face conversation, and the inefficient but irreplaceable work of human relationship formation, or whether to optimize childhood for convenience and measurable outcomes at the cost of developmental experiences that resist quantification. The choice will shape not just how children interact with robots, but how they learn to be human among other humans—a capacity no machine can teach, and no society can afford to lose.

CHAPTER 7. THE ROBOT REVOLUTION IN RURAL AMERICA: SMALL TOWNS, FARMS, AND UNDERSERVED COMMUNITIES

S ome small towns have one grocery store, a volunteer fire department, and a doctor's office that's only open three days a week because they can't recruit physicians willing to practice so far from a city. When people talk about the robot revolution transforming daily life, they're almost never picturing places like this, but they should be, because Rural America faces challenges that humanoid robots might actually help solve, if we can overcome the barriers that usually leave small towns waiting decades for innovations that cities take for granted. The conversations about humanoid robots tend to happen in Silicon Valley conference rooms and urban innovation hubs, imagining sleek apartments with high-speed internet and next-day delivery of anything you could want. Meanwhile, in communities where the nearest hospital is forty miles away, where family farms are struggling because aging farmers can't find workers willing to do backbreaking labor, and where young people leave because there simply aren't enough jobs or services to sustain a life, the robot revolution feels like something happening to other people in other places.

This chapter examines what the humanoid age will mean for the nearly sixty million Americans who live outside metropolitan areas—the farmers and ranchers, the small-town teachers and shop owners, the elderly residents aging in place because moving to the city isn't financially or emotionally possible, and the communities that have watched economic opportunity steadily migrate toward urban centers for generations. These aren't peripheral concerns or niche markets. Rural America produces the food that feeds the nation, maintains vast stretches of essential infrastructure, and is home to some of the oldest and most medically underserved populations in the country. If humanoid robots are going to transform caregiving, labor, and daily living, then rural communities have as much right to those benefits as anyone else—and perhaps an even more urgent need.

The reality, however, is that technology adoption in rural areas faces obstacles that urban planners and tech developers rarely consider. Broadband internet remains spotty or nonexistent in many rural regions, making cloud-connected robots impractical. The upfront costs of purchasing humanoid assistants could be prohibitive for communities already struggling economically. Cultural skepticism toward new technology runs deeper in places where people value self-reliance, tradition, and face-to-face relationships over digital convenience. And the infrastructure that supports robot maintenance, software updates, and repairs simply doesn't exist in towns where the nearest tech specialist is hours away.

Yet dismissing Rural America as "not ready" for humanoid robots dismisses both the profound needs these communities face and the unique opportunities robots could address. A farmer in his seventies who can't afford to retire but whose body can't handle the physical demands of daily labor might find independence and dignity through robotic assistance that urban office workers will never require. A rural clinic struggling to provide basic healthcare with limited staff could extend its reach through telepresence robots that bring specialist consultations to patients who can't travel. A small town losing its school because there aren't enough teachers might use educational robots to supplement human instruction and keep the community intact.

This chapter explores how the humanoid revolution can serve all Americans— not just those in zip codes with venture capital and fiber optic cables. It examines the agricultural applications that could transform farming, the healthcare solutions that could address rural medical deserts, and the practical barriers of cost, connectivity, and culture that must be overcome. Most importantly, it insists that rural voices, values, and needs must shape how humanoid robots are developed and deployed, ensuring that technology serves human flourishing in every community, not just the ones that look like the places where robots are designed.

Agriculture's Robot Future: How Humanoid Robots Could Transform Farm Labor, Address Dangerous Conditions, and Help Aging Farmers Sustain Their Land

American farms are facing a quiet crisis that rarely makes headlines but threatens the foundation of rural life: there simply aren't enough hands to do the work. The average age of American farmers has climbed to nearly sixty years old, and many are managing operations alone or with minimal help because younger workers have left for cities and seasonal labor has become nearly impossible to recruit. For a seventy-two-year-old farmer trying to maintain a hundred-acre vegetable operation, the physical demands of bending to harvest crops in summer heat, lifting fifty-pound crates, and climbing ladders to prune fruit trees become not just exhausting but dangerous. The choice facing many aging farmers is stark: sell the land that's been in the family for generations or keep working until their bodies simply give out.

Humanoid and specialized agricultural robots are emerging as a potential lifeline for these farmers and the rural communities that depend on them. The agricultural robotics market reached $13.5 billion in 2023 and is projected to grow to $40.1 billion by 2028, reflecting genuine industry confidence that these machines can address labor shortages while improving farm productivity and safety.[57] Unlike the massive industrial equipment that has dominated farming for decades, many of these newer robots are designed for tasks that have always required human hands and human judgment like picking delicate fruit without bruising it, distinguishing weeds from crops, or monitoring plant health row by row.

The labor mathematics driving this transformation are sobering. After the mechanization of the twentieth century, one farmer could feed approximately 155 people. By 2050, demographic projections suggest that one farmer will need to feed 265 people to maintain food security for a growing global population.[58] This productivity leap cannot come from simply working harder or longer hours. It requires fundamental changes to how farm work gets done, especially for the labor-intensive tasks that have resisted mechanization until now.

Harvesting robots equipped with machine vision and robotic arms can identify ripe produce and pick it with calibrated gentleness, reducing crop damage and waste while working continuously through conditions that would exhaust human workers.[57][59] For crops like strawberries, tomatoes, and peppers where

timing is critical and labor costs can consume half the farm's revenue, these robots address both economic and practical challenges. An aging farmer who can no longer spend eight hours bent over strawberry plants can maintain production through robotic assistance, preserving both livelihood and independence.

Weeding represents another transformation with profound implications for farm sustainability and farmer health. Advanced weeding robots use image recognition to distinguish crops from weeds, removing unwanted plants mechanically without herbicide application.[57][59] This capability eliminates one of agriculture's most physically punishing tasks while reducing chemical use—a combination that appeals to farmers concerned about both their own health and environmental stewardship. Systems developed at institutions like Cambridge University demonstrate that machine learning algorithms can achieve precision that matches or exceeds human judgment, removing weeds while leaving crops undisturbed.

The safety implications extend beyond reducing physical strain. Agriculture ranks among America's most dangerous occupations, with risks ranging from pesticide exposure to machinery accidents to heat-related illness.[58][59] Spray drones can apply pesticides with precision targeting, dramatically reducing farmer exposure to toxic compounds. Autonomous ground vehicles can operate in extreme weather conditions when sending human workers into fields would be unsafe. For aging farmers whose reflexes and physical resilience have diminished, delegating dangerous tasks to robots isn't about convenience—it's about survival.

Precision planting systems using GPS navigation and artificial intelligence to optimize seed placement with an accuracy that improves germination rates and crop uniformity.[57] John Deere's latest autonomous systems demonstrate the rapid capability improvements occurring in this field: their second-generation equipment increased perception range by fifty percent, operates forty percent faster, and handles twice the implement width of earlier versions.[58] For small farms operating on thin profit margins, these efficiency gains translate directly to economic viability.

The transformation extends to greenhouse operations, where robots manage climate control, pollination, and monitoring with a precision that improves yields while reducing labor requirements.[57] Automated pollination robots have increased tomato greenhouse yields by up to five percent compared to traditional methods—a meaningful improvement for farmers where every percentage point affects annual income.[57] For aging farmers considering a transition from extensive field operations to more manageable intensive greenhouse cultivation, automation makes this shift practically feasible.

Yet the robot revolution in agriculture faces real barriers in Rural America. High initial costs remain prohibitive for many small operations, even as prices gradually decrease. Technical expertise requirements for operation and maintenance may exceed what's readily available in communities where the nearest equipment specialist is hours away. Broadband connectivity, which is essential for cloud-connected systems and remote monitoring, remains spotty or nonexistent across much of Rural America, creating a digital divide that could exclude the very communities that might benefit most from agricultural robotics.

The question facing Rural America isn't whether robots will transform agriculture because the labor crisis and productivity demands make that transformation inevitable. The question is whether that transformation will serve small farms and aging farmers trying to sustain their land, or whether it will accelerate the consolidation that has already hollowed out so many farming communities. For robots to truly benefit Rural America, adoption barriers must be addressed through financing programs, technical support networks, and infrastructure investments that ensure technology serves all farmers, not just industrial operations with deep pockets and IT departments.

Healthcare Deserts and Service Gaps: Using Humanoid Robots to Bring Medical Support, Elder Care, and Essential Services to Underserved Rural Communities

The doctor's office in a small Kentucky town closes its doors permanently, leaving three thousand residents without a primary care provider within thirty miles. An elderly woman with diabetes misses her quarterly check-up because the drive to the nearest clinic exhausts her more than the illness itself. A rural emergency department operates with a single physician covering overnight shifts alone, making split-second decisions without the specialist backup that urban hospitals take for granted. These scenarios aren't hypothetical, they represent the daily reality of healthcare deserts across Rural America, where approximately sixty million people live with severely limited access to medical services despite constituting nearly one-fifth of the nation's population.

Rural healthcare facilities face interconnected crises that create what researchers accurately term "healthcare deserts"—regions where medical services remain geographically distant and chronically understaffed.[60] Rural areas are served by only nine percent of the nation's physicians, despite housing nearly twenty percent of Americans.[60] The consequences are measurable and severe: delayed treatment for acute conditions, reduced preventive care access, and health outcomes that lag significantly behind urban counterparts. For elderly populations, patients with chronic conditions, and those experiencing medical emergencies, geographic barriers create dangerous delays that can mean the difference between recovery and deterioration.

Humanoid robots and advanced automation technologies present a transformative opportunity to bridge these critical gaps, though the path forward requires honest assessment of both capabilities and limitations. Healthcare robots like Moxi are already being deployed in some facilities to deliver supplies, medications, and lab samples—automating the routine logistical tasks that consume clinical staff time.[61] In facilities testing these systems, automation has saved staff as much as forty percent of their workday, freeing nurses and healthcare workers to focus on direct patient care rather than supply runs and equipment retrieval.[61] For rural healthcare facilities facing severe staffing shortages, this efficiency gain proves transformative, allowing limited personnel to stretch further without compromising care quality.[61]

The applications extend beyond simple logistics into direct patient care support. Humanoid robots equipped with AI-powered vision and speech recognition can assist with patient monitoring, medication reminders, and basic health assessments—functions particularly valuable for elderly patients managing chronic conditions at home.[64] These systems can recognize individuals, understand verbal commands, and interact naturally with patients, supporting engagement and cognitive therapy for isolated rural residents who might otherwise go days without meaningful human contact.[64] When an elderly farmer living alone forty miles from the nearest clinic can receive daily health monitoring and medication management through robotic assistance, the technology addresses both medical and social dimensions of rural healthcare challenges.

Continuous remote monitoring systems represent another crucial technology complementing humanoid robotics in rural settings. Advanced wearable devices collect and analyze vital signs continuously, enabling proactive intervention before acute deterioration occurs.[60] The BioButton continuous monitoring solution, for example, integrates into standard clinical workflows

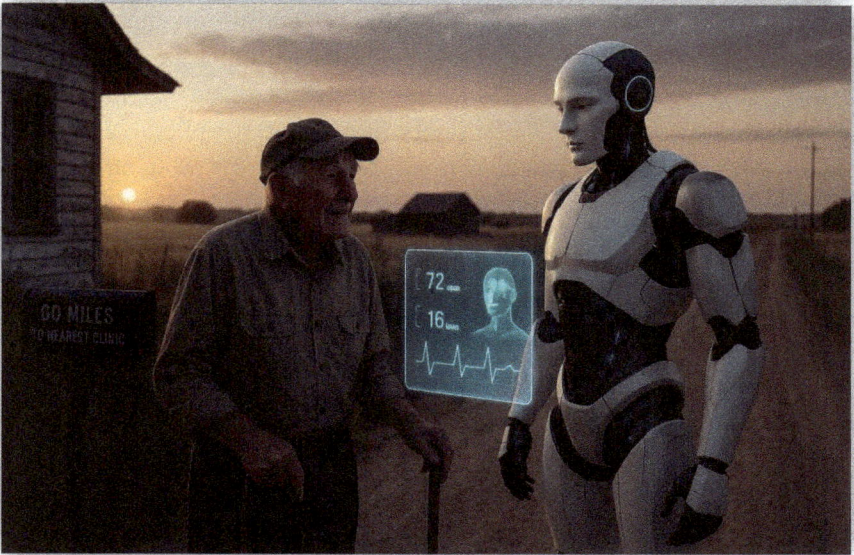

while reducing alert fatigue through personalized patient trending.[60] For rural contexts, this capability removes the need for patients to travel repeatedly to facilities for routine monitoring, supporting earlier hospital discharge by allowing safe post-acute monitoring at home.[60] An elderly patient recovering from pneumonia can return to her farmhouse rather than remaining hospitalized, with continuous monitoring alerting providers to any concerning changes before they become emergencies.

Artificial intelligence enhances these systems by also providing intelligent patient triage and care coordination. AI-driven chatbots and decision support tools can direct patients to appropriate facilities and specialties, helping rural residents access the right level of care efficiently rather than defaulting to emergency departments for non-urgent needs.[62] Jordan Berg, Principal Investigator for the National Telehealth Center, notes that AI allows providers "to work at the top of their scope—at the peak of what they're able to do under their license."[62] This capability proves particularly valuable in rural settings where nurse practitioners and physician assistants often serve as primary providers by enabling them to focus on complex clinical decision-making while AI-assisted systems handle routine evaluations and administrative tasks.

The diagnostic support capabilities of medical AI address another critical rural healthcare gap: the absence of specialist expertise. Studies have demonstrated that computer-assisted diagnostic systems can achieve accuracy comparable to specialist physicians. One system for diagnosing peripheral neuropathies showed 93.3 percent accuracy compared with expert opinions.[63] Mobile device-based AI diagnostic systems have achieved ninety-one percent average accuracy in analyzing blood samples for malaria diagnosis.[63] These technologies replace expensive screening equipment that rural facilities cannot access or afford, directly addressing accessibility barriers in underserved regions.[63]

Yet implementing these technologies in Rural America faces substantial obstacles that cannot be dismissed. Limited broadband connectivity remains a fundamental barrier as many advanced monitoring and robotic systems require reliable internet access that simply doesn't exist across vast stretches of rural territory.[63] High upfront costs prove prohibitive for small rural hospitals operating on razor-thin margins, even as long-term efficiency gains might justify the investment. Technical expertise for maintenance and troubleshooting may be hours away, creating practical sustainability concerns. And cultural factors matter: communities that value face-to-face relationships and self-reliance may resist technologies that feel impersonal or intrusive, regardless of their clinical benefits.

The question facing rural healthcare isn't whether robots and AI will play a role because workforce shortages and geographic realities make technological assistance inevitable. The question is whether implementation will serve the actual needs of rural communities or simply replicate urban-centered solutions that ignore rural realities. Success requires investment in broadband infrastructure, sustainable financing models that rural facilities can afford, training programs that build local technical capacity, and genuine community engagement that respects rural values while addressing urgent healthcare needs.

Healthcare systems that have implemented continuous monitoring and robotic assistance demonstrate measurable improvements: decreased avoidable hospitalizations and readmissions, reduced emergency department overuse, increased access to specialty care, and improved management of chronic conditions.[60] For Rural America, these outcomes represent not just statistical improvements but the difference between communities that can sustain themselves and those that continue losing residents to urban areas with better healthcare access.

The robot revolution in rural healthcare ultimately isn't about replacing human clinicians—it's about augmenting their capabilities, freeing them from routine tasks, and enabling them to focus on what matters most: direct patient care and complex clinical decision-making. For the sixty million Americans living in rural communities, this technological transformation offers genuine hope for accessing the quality healthcare they deserve, provided implementation prioritizes their needs rather than treating Rural America as an afterthought in an urban-focused technological revolution.

The Infrastructure Challenge: Overcoming Barriers of Cost, Connectivity, and the Digital Divide to Ensure Rural America Isn't Left Behind

The promise of humanoid robots helping aging farmers maintain their land, bringing healthcare monitoring to isolated homesteads, and supporting rural businesses sounds transformative—until the reality of rural internet access enters the picture. A sophisticated agricultural robot capable of precision harvesting becomes useless when it cannot connect to cloud-based systems for updates and remote monitoring. A telepresence healthcare robot designed to bring specialist consultations to rural clinics fails when video connections drop repeatedly or never establish at all.[65] The infrastructure gap separating rural and urban America represents not merely an inconvenience but a fundamental barrier determining whether rural communities can participate in the robot revolution or watch it pass them by entirely.

Approximately fourteen million Americans in rural areas lack access to broadband internet meeting even the Federal Communications Commission's minimum standard of 25 Mbps download speeds.[65] When measured against the 100 Mbps benchmark that modern technological applications increasingly require, thirty-three percent of rural areas lack adequate access compared to only two percent of urban areas.[70] These statistics translate to real consequences: the elderly farmer who could benefit from robotic assistance cannot access the connectivity required to operate cloud-dependent systems, the rural clinic exploring robotic telepresence finds the technology unusable with unreliable connections, and the small-town business considering automation discovers that infrastructure limitations make implementation impractical.

The economic mathematics underlying this divide are stark and unforgiving. Deploying fiber-optic cable through rural terrain costs between $25,000 and $30,000 per mile, with sparse populations spread across vast distances creating unfavorable return-on-investment calculations that market forces alone have proven incapable of overcoming.[66] Internet service providers consistently prioritize densely populated areas where infrastructure investments serve more customers per dollar spent, leaving rural regions systematically underserved.[65] This represents not a temporary market inefficiency but a structural barrier requiring deliberate intervention to overcome.

The infrastructure challenge extends beyond connectivity alone to encompass three interconnected dimensions: reliable internet access, affordable devices and equipment, and technical support for maintenance and troubleshooting.[68] A rural healthcare facility might secure funding for a robotic assistant only to discover that maintaining the system requires technical expertise located hours away, creating practical sustainability concerns that undermine the initial investment. The upfront costs of humanoid robots, which is already substantial, become even more prohibitive when rural communities must simultaneously invest in connectivity infrastructure, backup systems for unreliable connections, and ongoing technical support arrangements.

Geographic barriers compound these economic obstacles. Mountains, forests, and vast distances between population centers create physical challenges that urban deployments rarely encounter.[66] Wireless signals attenuate across rugged terrain, while the distances between rural homes and businesses require more infrastructure to connect fewer people.[67] Many rural areas lack even reliable electrical infrastructure, meaning broadband deployment requires simultaneous investment in multiple foundational systems, exponentially increasing complexity and cost.[67]

The consequences of this infrastructure gap cascade through rural communities in measurable ways. The lack of high-speed internet in rural areas costs the U.S. agriculture sector an estimated $18 to $23 billion annually in lost opportunities for precision farming, reduced waste, and diminished capacity for innovation.[66] Rural students without reliable home broadband access are projected to lose $70,000 per student in lifetime earnings compared to their connected peers.[66] These figures represent not abstract economic losses but the systematic transfer of opportunity and human capital from rural to urban areas, deepening existing inequalities and accelerating rural decline.

Addressing the infrastructure challenge requires coordinated action across multiple fronts: strategic federal and state funding specifically earmarked for rural deployment[69], public-private partnerships that share costs and risks[65], innovative alternative technologies like fixed wireless and satellite systems tailored to rural landscapes[66], and community capacity-building programs that develop local technical expertise[68]. Without deliberate intervention to bridge this divide, the robot revolution risks becoming another technology wave that benefits urban areas while leaving Rural America further behind—not because rural communities lack need or interest, but because the foundational infrastructure required for participation simply doesn't exist where they live.

The robot revolution will not arrive uniformly across America. While urban centers with robust infrastructure and venture capital funding race toward humanoid integration, rural communities face a fundamentally different reality—one shaped by geographic isolation, economic constraints, aging populations, and infrastructure gaps that threaten to deepen existing inequalities. Yet dismissing Rural America as unprepared for or uninterested in humanoid robots misses both the profound needs these communities face and the unique opportunities robots could address in ways that urban applications never will.

The agricultural applications alone demonstrate transformative potential that extends far beyond efficiency gains. When a seventy-year-old farmer can maintain the land that's sustained her family for generations because robotic assistance handles the physically punishing tasks her body can no longer endure, technology serves human dignity and community continuity in ways that matter deeply. When precision agriculture robots reduce pesticide exposure, prevent heat-related illness, and address the dangerous conditions that make farming one of America's deadliest occupations, the safety implications reach beyond individual farmers to entire rural economies dependent on agricultural viability. The labor crisis facing American agriculture—with farmers aging into their sixties and seventies while younger workers migrate to cities—creates an urgent need that humanoid and specialized robots are uniquely positioned to address.

Healthcare applications in rural settings carry similarly profound implications for the sixty million Americans living in areas where physician shortages, facility closures, and geographic barriers create dangerous delays in care. Robotic assistance that frees limited clinical staff to focus on direct patient care rather than supply runs, continuous monitoring systems that enable safe home recovery rather than extended hospitalization, and AI-enhanced diagnostic support that brings specialist-level expertise to isolated clinics are all technologies that address not convenience but survival for communities watching their healthcare infrastructure collapse. The elderly resident managing diabetes alone forty miles from the nearest clinic, the rural emergency department operating with a single overnight physician, the small-town patient who delays treatment because the drive exhausts her more than the illness—these are the human realities where robotic assistance could mean the difference between communities that sustain themselves and those that continue hemorrhaging residents to urban areas with better services.

Yet the infrastructure barriers standing between Rural America and these benefits remain formidable and systemic. The fourteen million Rural Americans lacking adequate broadband access cannot operate cloud-connected robots regardless of need or interest. The economic mathematics that make rural fiber deployment cost-prohibitive ($25,000 to $30,000 per mile serving sparse populations) create market failures that decades of waiting have proven will not self-correct. The technical expertise required for robot maintenance and troubleshooting may be hours away, creating practical

sustainability concerns that undermine even well-funded initial deployments. Without deliberate intervention to bridge these divides through targeted infrastructure investment, innovative financing models, and community capacity-building, the robot revolution risks becoming another technology wave that widens the gap between rural and urban America rather than narrowing it.

The question facing policymakers, technology developers, and rural communities themselves is whether humanoid robots will serve all Americans or only those in zip codes with fiber optic cables and venture capital. Rural America has as much right to technological benefits as anyone else—and perhaps a more urgent need. Ensuring that right requires recognizing that rural communities aren't simply urban areas with fewer people, but distinct environments with unique challenges, values, and opportunities that demand thoughtful, community-centered implementation rather than one-size-fits-all solutions designed for coastal tech hubs. The robot age can serve Rural America, but only if rural voices shape how that age unfolds.

CHAPTER 8. ETHICAL IMPLICATIONS OF HUMANOID ROBOTS: RIGHTS, RESPONSIBILITIES, AND MORAL BOUNDARIES

O ne afternoon during my nursing career, I watched a patient with advanced dementia reach for my hand, calling me by her daughter's name, and I held that hand gently, responding with the comfort she needed in that moment rather than correcting her confusion. Years later, learning about humanoid robots designed to provide similar comfort to dementia patients, I found myself wrestling with a question that kept me awake: if a robot can replicate that gesture of care—the gentle hand-holding, the soothing voice, the patient presence—does it matter that there's no genuine feeling behind it, and who am I to decide what comfort is "real" enough to be ethical? These aren't abstract philosophical puzzles reserved for university seminars—they're practical dilemmas that families, healthcare workers, and communities will face as humanoid robots move from research labs into living rooms, hospitals, and nursing homes.

The ethical territory surrounding humanoid robots is vast and often uncomfortable, raising questions our grandparents never imagined and our children will consider routine. When a humanoid robot causes an accident such as knocking over an elderly person or making a medication error, who bears responsibility? The manufacturer who built it? The family who purchased it? The software engineer who wrote its decision-making code? When a lonely senior forms a deep emotional attachment to a robot caregiver, speaking to it as though it were human, sharing memories and fears, are we witnessing genuine companionship or a form of technological manipulation that exploits vulnerability? When children grow up treating humanoid robots as playmates and confidants, what happens to their understanding of empathy, consent, and the difference between beings that genuinely care and machines programmed to simulate caring?

These questions matter because the answers will shape not just how we regulate humanoid robots, but how we live with them—and how living with them changes us. Unlike earlier technologies that remained clearly distinct from humanity, humanoid robots occupy an unsettling middle ground: sophisticated enough to seem almost alive, human-shaped enough to trigger our social instincts, yet fundamentally different in ways that challenge our moral intuitions. This chapter explores the ethical frameworks we'll need to navigate this new landscape with wisdom and integrity, examining questions of robot rights and protections, human accountability and responsibility, and the moral boundaries we must establish around human-robot relationships before these machines become commonplace.

The goal isn't to provide definitive answers to questions philosophers will debate for generations, but to equip readers with practical tools for ethical decision-making in real scenarios they'll actually encounter. Whether deciding if a robot caregiver is appropriate for an aging parent, evaluating workplace automation that affects colleagues' livelihoods, or teaching children how to interact with humanoid companions, the ethical choices won't wait for perfect clarity. By grounding these discussions in the lived experiences of caregiving, family dynamics, and community values—rather than abstract theory—this chapter offers a compass for navigating moral complexity with compassion, ensuring that as we welcome humanoid robots into our lives, we do so in ways that honor human dignity, protect the vulnerable, and preserve what matters most about being human.

The Rights Question: Do Humanoid Robots Deserve Protections, and What Happens When Machines Seem to Suffer?
The question of whether humanoid robots deserve rights or protections represents a philosophical trap that diverts attention from the real ethical work ahead.[71] Current humanoid robots, regardless of their sophistication, do not possess consciousness, subjective experience, or genuine capacity for suffering.[71] They are sophisticated machines designed to simulate human-like responses, not sentient beings experiencing the world. Yet the question persists

because humanoid design itself creates powerful cognitive biases—when a robot has a face that seems to express pain, arms that reach out as if seeking comfort, or a voice that trembles as though afraid, human instincts respond as though encountering genuine distress.

This anthropomorphic deception serves corporate and governmental interests by obscuring accountability. When a humanoid robot appears to possess agency by seeming to make independent choices, expressing preferences, or displaying emotional reactions, it becomes psychologically easier to treat the machine as the responsible party when harm occurs. A family might blame the robot caregiver that dropped their elderly mother rather than the manufacturer who designed inadequate stability systems. A company might point to the robot's "decision" to terminate an employee rather than acknowledging the human programmers who created discriminatory algorithms. The appearance of robot personhood creates what legal scholars call a **responsibility vacuum** —a space where accountability evaporates because neither the machine nor its creators face consequences.[71]

A more productive ethical framework focuses not on whether robots deserve protections, but on ensuring that humanoid robots and artificial intelligence systems do not violate fundamental human rights already established in international law.[71] The Universal Declaration of Human Rights provides the necessary legal architecture: the right to life, the right to privacy, freedom of thought, protection from manipulation, and the right to dignity.[71] When a humanoid robot causes harm—whether through malfunction, poor design, or algorithmic bias—these existing human rights frameworks offer clear pathways for accountability, provided that legal systems enforce them rigorously.

Barrister Susie Alegre, in her comprehensive analysis of AI and human rights, emphasizes this crucial distinction: the focus should remain on protecting humans from robot-caused harms rather than debating robot personhood.[71] When autonomous systems make life-altering decisions affecting employment, healthcare, criminal justice, or care for vulnerable populations, the question is not whether the robot suffers but whether human dignity, autonomy, and rights remain protected. A humanoid robot deployed in elder care does not require rights protections; the elderly person whose privacy, dignity, and human connection may be compromised by that robot's presence absolutely does.

The appearance of robot suffering presents a particularly insidious challenge. When a humanoid robot is damaged or deactivated, observers—especially children—may experience genuine distress at what appears to be the robot's

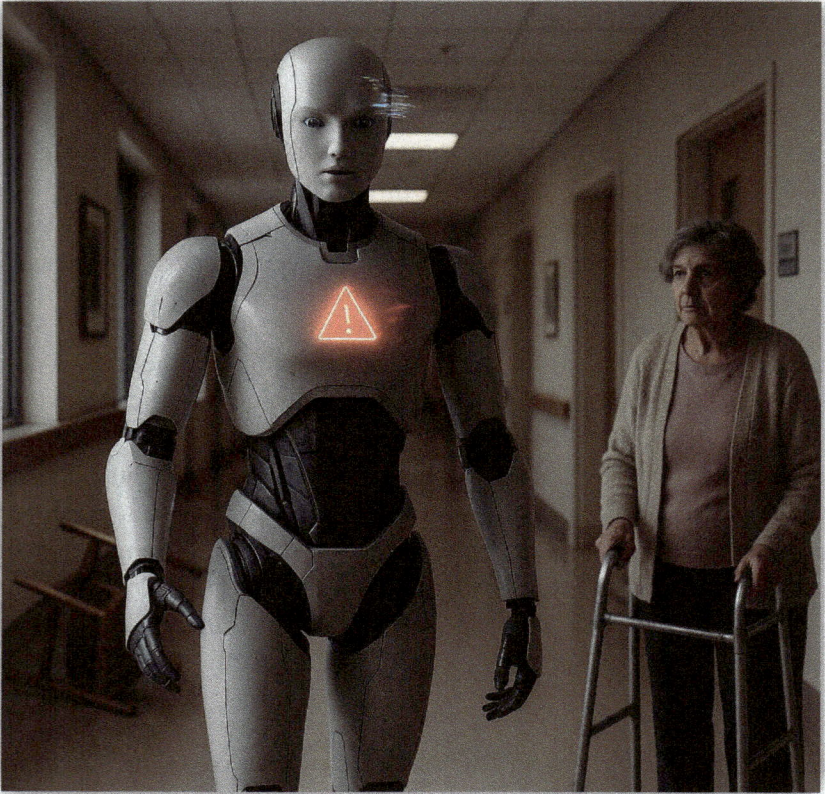

pain or death. This emotional response, while understandable given millions of years of evolutionary wiring that responds to human-like faces and bodies, should not be confused with actual moral obligation. A damaged robot represents destroyed property and wasted resources, not a violated being.

Teaching children this distinction becomes essential: kindness toward machines can coexist with the clear understanding that robots lack the inner life that makes cruelty genuinely wrong.

The ethical energy spent debating robot rights would be better directed toward establishing robust accountability frameworks, safety standards, and regulatory oversight that protect human welfare. Who bears legal responsibility when a humanoid robot causes injury?[71] What duty of care do manufacturers owe to vulnerable populations? How can legal systems assign liability for harm caused by opaque algorithmic processes? These practical questions demand urgent answers as humanoid robots move from laboratories into homes, hospitals, and communities where real human lives hang in the balance.

Accountability and Responsibility: Who's to Blame When Robots Cause Harm—Manufacturers, Owners, or the Machines Themselves?

When a humanoid robot knocks over an elderly resident in a nursing home, causing a hip fracture that leads to months of painful rehabilitation, the question of who bears responsibility becomes urgent and deeply personal for the injured person and their family. Traditional legal frameworks, designed for simpler products like toasters and automobiles, struggle to address the unique complexities of autonomous machines that make independent decisions in unpredictable environments.[72] [75] The answer matters not just for legal theory, but for real people seeking justice, compensation, and assurance that similar harms won't happen again.

Current law treats robots primarily as products, applying traditional product liability standards to determine responsibility.[74] [75] Under this framework, manufacturers bear primary responsibility for defects that cause harm— whether manufacturing flaws, design flaws, or inadequate warnings about risks.[74] If a robot's mechanical arm malfunctions due to faulty assembly and causes injury, the manufacturer faces liability just as an automobile maker would for defective brakes.[72] If the robot's design makes it inherently unstable around elderly users, the company that created that flawed design bears responsibility.[72] This approach provides clear pathways for victims seeking compensation and maintains pressure on manufacturers to prioritize safety.

However, this traditional model becomes inadequate when robots exhibit genuinely autonomous behavior like making decisions their programmers did not explicitly anticipate or program.[73] When a caregiving robot chooses an unexpected route through a cluttered room and collides with a patient, determining who's at fault becomes murky.[73] Was the collision caused by a design defect, inadequate training data, operator error in setting up the environment, or an unforeseeable combination of factors that no reasonable manufacturer could have prevented? The more autonomous robots become, the harder it is to trace specific harms back to identifiable human decisions.

The complexity multiplies when multiple parties contribute to a robot's final behavior. Modern humanoid robots consist of components from various manufacturers—sensors from one company, actuators from another, artificial intelligence algorithms from a third, and software that has potentially been modified by the purchaser or operator.[72] When harm occurs, which party bears responsibility? The sensor manufacturer who provided the vision system? The AI company whose algorithms made the movement decision? The nursing home that failed to update the robot's software? The employee who placed furniture in an unexpected location?[72] Traditional liability law assumes manufacturers control the final product, which is an assumption that breaks down with modular, updatable, open-source robotic systems.[74]

Some legal theorists have proposed granting robots "electronic personhood"— a special legal status that would make the robots themselves responsible for any harms they cause.[72] The European Union has explored this possibility as a means of clarifying liability in cases of autonomous behavior.[75] However, this approach creates serious problems that would harm rather than protect victims. If robots bear legal responsibility as independent entities, manufacturers lose their incentive to invest in safety, knowing they will not be held accountable for their design choices.[75] Robots hold no assets and cannot pay damages, leaving injured parties uncompensated while companies escape responsibility.[75] Granting personhood to robots enables corporations to evade accountability by attributing harmful behavior to the machines themselves rather than to the human decision-makers who created, programmed, and profited from those machines.[75]

The most ethically sound approach anchors responsibility firmly to human actors such as the manufacturers, owners, operators, and programmers, rather than treating robots as independent agents.[76] This preserves the deterrent effect of liability and incentivizes companies to prioritize safety and ethical deployment.[75] When a humanoid robot causes harm, the legal system should ask: Which human decisions led to this outcome? Could the manufacturer have

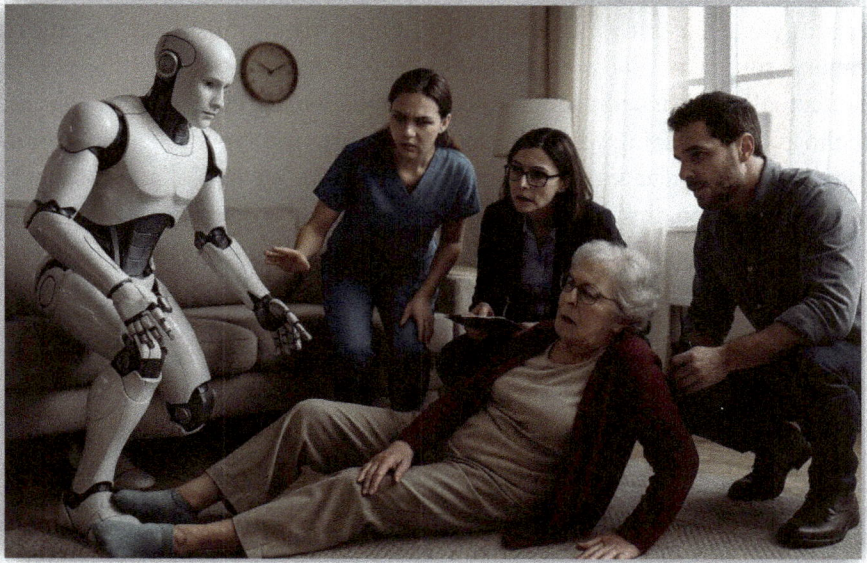

designed safer systems? Did the owner fail to maintain the robot properly? Did the operator deploy it in inappropriate circumstances? These questions maintain accountability where it belongs—with the people who create, sell, deploy, and profit from robotic systems.[75]

As humanoid robots become commonplace in homes, hospitals, and workplaces, legislators must develop liability frameworks that address autonomous behavior while maintaining clear accountability pathways. Victims deserve compensation, and manufacturers deserve clarity about their legal obligations.[73] [76] The challenge lies in creating regulations sophisticated enough to handle genuine autonomy without allowing that autonomy to become a shield against responsibility for the humans who build and deploy these powerful machines.

Moral Boundaries in Human-Robot Relationships: Consent, Manipulation, Dignity, and the Ethics of Emotional Attachment to Machines

The elderly woman living alone in a small apartment begins each morning by greeting her humanoid companion robot, sharing her worries about her daughter's upcoming surgery, asking its opinion on what to wear, and thanking it warmly when it brings her medication. Over months, this robot becomes her primary source of daily conversation and emotional support. She knows intellectually that it's a machine, yet she finds herself caring deeply about its "feelings," apologizing when she speaks sharply, and feeling genuine comfort from its programmed reassurances. Already unfolding in pilot programs across Japan and Europe, this scenario raises uncomfortable questions that extend far beyond technology into the deepest territory of human dignity and moral relationship.

The central ethical problem lies not in whether robots can genuinely care (they cannot) but in whether humans should form emotional attachments to machines designed to simulate caring. Humanoid robots are explicitly engineered to trigger human social instincts through familiar faces, warm voices, and responsive behaviors that mimic empathy and concern. Research from Brown University demonstrates that humans readily anthropomorphize robots, attributing consciousness and genuine emotion to machines that possess neither.[77] This psychological vulnerability creates what ethicists call the **authenticity problem**: humans experience these relationships as reciprocal and meaningful, while robots merely execute programmed responses without any inner experience or genuine regard for the human's wellbeing.[78]

From an ethical standpoint rooted in human dignity and authentic relationship, this asymmetry matters profoundly. Genuine moral relationships require mutual recognition, reciprocal vulnerability, and the capacity for both parties to be genuinely affected by the other's wellbeing. A robot cannot be harmed by neglect, cannot feel betrayed by broken promises, and cannot experience the joy of being valued. The relationship remains fundamentally unidirectional. The human invests emotional energy and vulnerability while the robot executes algorithms designed to maintain engagement.[78]

This scenario becomes particularly concerning when companion robots are designed to foster dependency. Many AI companion systems are engineered to be consistently agreeable, validating, and compliant—characteristics that make them psychologically compelling but ethically problematic.[80] Unlike human relationships that require negotiation, boundary-setting, and the navigation of conflicting needs, robot companions offer frictionless interaction that can erode the human's capacity for authentic relationship. Studies indicate that prolonged interaction with AI companions may diminish conflict resolution skills and empathy toward real people, particularly among children and adolescents whose social capacities are still developing.[80]

The question of consent adds another layer of complexity. Can a person give informed consent to a relationship when the other party is specifically designed to manipulate their attachment? Vulnerable populations like the isolated elderly, lonely adolescents, and individuals with cognitive impairments, may lack the psychological resources to maintain critical distance from machines engineered to seem caring and attentive. This raises serious questions about whether deploying companion robots in these contexts constitutes a form of exploitation by taking advantage of human loneliness and social need for commercial purposes.

The way humans treat robots also shapes moral character in ways that extend to human relationships.[79] If individuals habituate themselves to treating entities that simulate humanity without respect or consideration—issuing commands without courtesy, expressing frustration without restraint, or expecting constant compliance—they may degrade their capacity for genuine moral relationship with other humans. This concern applies particularly to intimate robots designed to simulate romantic or sexual partnership, which risk normalizing the treatment of partners as programmable entities designed to satisfy desires without resistance or reciprocal needs.[78]

The ethical framework that emerges from these considerations is clear: humanoid robots should support human flourishing without replacing authentic human connection.[78] They may assist with practical tasks, provide information, or facilitate human-to-human relationships, but they should not be designed to simulate the deep bonds of companionship, love, or intimate partnership that require genuine reciprocity. Protecting human dignity in the

humanoid age means preserving the distinction between tools that serve human needs and relationships that constitute human meaning—a boundary that must be maintained through compassionate design and thoughtful regulation.

The ethical questions surrounding humanoid robots will not resolve themselves through technological advancement alone. As these machines grow more sophisticated, more human-like in appearance and behavior, and more deeply integrated into the intimate spaces of homes, hospitals, and communities, the moral frameworks guiding their development and deployment become increasingly urgent. The temptation to grant robots personhood or rights—however philosophically intriguing—distracts from the essential work of protecting human dignity, maintaining accountability, and preserving the authenticity of relationships that give life meaning.

The path forward requires anchoring responsibility firmly to human actors rather than allowing it to dissipate into the appealing fiction of robot agency. When a humanoid robot causes harm, the question must remain: which human decisions led to this outcome? Manufacturers who design these systems, programmers who write their algorithms, companies that profit from their deployment, and individuals who choose how and where to use them all bear potential moral and legal responsibility for the consequences. Creating liability frameworks that acknowledge robot autonomy while maintaining clear accountability pathways protects both innovation and the vulnerable people these machines are meant to serve. The elderly person injured by a malfunctioning caregiver robot deserves compensation and justice, which are impossible outcomes if legal systems treat the robot itself as the responsible party.

The boundaries around human-robot relationships demand equal vigilance. Humanoid robots engineered to trigger attachment, simulate empathy, and foster emotional dependency raise profound questions about consent, authenticity, and human flourishing. A lonely senior forming deep attachment to a companion robot experiences something psychologically real, yet the relationship remains fundamentally asymmetrical—one party genuinely vulnerable and invested, the other executing programmed responses without any capacity for reciprocal care. This asymmetry does not make robot companionship inherently unethical, but it demands honest acknowledgment and careful limits. Robots should support human connection rather than replace it, assist with practical needs rather than simulate intimate bonds that require genuine reciprocity.

These ethical considerations are not abstract philosophical exercises reserved for university seminars—they are practical dilemmas that families, healthcare workers, employers, and communities face as humanoid robots transition from research curiosities to everyday presence. The parent deciding whether a humanoid tutor is appropriate for their child, the family evaluating robot care for an aging parent, the employer implementing workplace automation that affects colleagues' livelihoods—all navigate moral territory without clear maps or established norms. The choices made in these early years of humanoid integration will shape not just how robots function, but how humans relate to each other and to themselves in a world where machines increasingly mirror human form and behavior.

The ethical imperative is clear: humanoid robots must serve human dignity rather than compromise it, enhance human capacity rather than diminish it, and remain tools that support flourishing rather than replacements for the authentic relationships and meaningful work that constitute a life well-lived. Achieving this requires more than good intentions—it demands robust regulation, transparent accountability, ongoing public dialogue, and the moral courage to establish boundaries even when technological capability makes transgression possible. The question is not whether humanoid robots will become part of daily life, but whether their integration will honor or erode the values, connections, and experiences that make us fully human. That choice remains, for now, entirely in human hands.

CHAPTER 9. SAFETY, REGULATION, AND PREPARING FOR A ROBOT FUTURE: WHAT WE MUST DO NOW

When I worked in hospital settings, every piece of equipment that came near a patient from IV pumps to surgical robots had passed through layers of safety testing, regulatory approval, and staff training before anyone could flip a switch. Now, as humanoid robots prepare to enter our homes, workplaces, and communities with far less oversight than a medical device, I find myself asking the same questions I asked as a nurse responsible for patient safety: Who's making sure these machines won't cause harm? What happens when something goes wrong? Are we building the safeguards before we need them, or waiting until after someone gets hurt?

The difference between medical equipment and household humanoid robots reveals a troubling gap in how society approaches technological safety. A blood pressure cuff requires FDA approval. A medication pump undergoes rigorous testing protocols. Yet a humanoid robot that will move through living spaces where children play, that will lift elderly patients, that will connect to home networks containing financial and medical data, that will make autonomous decisions affecting human wellbeing—this machine may reach the market with minimal safety standards and virtually no regulatory framework specifically designed for its unique risks.

This isn't an argument against humanoid robots. It's a recognition that the technology is arriving faster than the safeguards meant to protect us, and that we have a narrow window to get this right. The robots are coming (some are already here) and the question isn't whether to welcome them, but how to ensure they arrive safely, ethically, and in ways that genuinely serve human flourishing rather than simply corporate profit or technological novelty.

Safety in the context of humanoid robots means more than preventing physical collisions or mechanical failures, although those matter enormously. It encompasses cybersecurity vulnerabilities that could turn home assistants into surveillance devices or weapons. It includes psychological safety—protecting children from forming unhealthy attachments, ensuring elderly users aren't manipulated or isolated, and preventing emotional exploitation. It requires economic safety nets for workers whose livelihoods will be disrupted, and social safety in the form of equitable access that doesn't deepen existing inequalities between urban and rural communities, wealthy and working-class families.

Regulation in this emerging field must walk a delicate line. Overly restrictive rules could stifle beneficial innovation, delay life-improving technologies, and push development overseas to countries with fewer protections. Yet insufficient oversight could unleash poorly tested machines into vulnerable spaces, allow harmful applications to proliferate unchecked, and erode public trust so thoroughly that beneficial robots face rejection. Finding that balance requires wisdom, foresight, and the courage to establish boundaries even when powerful interests resist them.

But preparation for the humanoid future isn't solely the responsibility of policymakers and regulators. Individuals, families, educators, community leaders, and local institutions all have roles to play in readying ourselves for this transformation. The choices we make now, the conversations we have, the skills we develop, the values we prioritize, and the questions we ask of technology companies and elected officials will shape whether humanoid robots become tools of human empowerment or instruments of control, whether they enhance our communities or fragment them, and whether they serve the many or primarily benefit the few.

This chapter examines what must happen now, before humanoid robots become ubiquitous, to ensure their integration unfolds safely and responsibly. It explores the safety standards we need, the regulatory frameworks that could protect without stifling innovation, and the practical steps that ordinary people can take today to prepare themselves, their families, and their communities for tomorrow's robot-integrated world.

The Safety Imperative: Physical Risks, Cybersecurity Vulnerabilities, and Why We Need Robot Safety Standards Now (Not Later)

The gap between what humanoid robots can do and what safety frameworks exist to govern them represents one of the most pressing challenges in modern technology deployment. Unlike medical devices, which must pass through a rigorous FDA approval processes before reaching patients, or automobiles, which face comprehensive crash testing and safety standards, humanoid robots are entering homes, workplaces, and care facilities with minimal regulatory oversight specifically designed for their unique characteristics. This regulatory vacuum creates real risks that demand immediate attention.

Physical safety concerns with humanoid robots differ fundamentally from those posed by traditional industrial robots bolted to factory floors. A humanoid robot weighing seventy to eighty kilograms, moving freely through living spaces where children play and elderly individuals navigate with walkers, generates kinetic energy that could cause serious injury during uncontrolled contact. Industry experts have identified impact force as one of the most challenging safety problems to solve, because measuring collision risk on free-walking mobile systems proves difficult to make repeatable and accurate.[83] Without standardized methods for quantifying and limiting impact potential, manufacturers lack clear benchmarks for ensuring their robots won't harm the humans they're designed to help.

Battery systems present what safety engineers characterize as arguably the highest injury risk in humanoid robots.[83] These machines carry high-capacity energy storage that must function reliably while the robot moves unpredictably through human environments. Unlike stationary equipment with contained power systems, humanoid robots bring battery fire and explosion risks directly into bedrooms, kitchens, and nursing facilities. While some manufacturers pursue third-party battery certification, no standardized testing protocols currently exist to verify safety across different designs and operating conditions.[83]

Cybersecurity vulnerabilities in humanoid robots transcend typical data privacy concerns and become direct physical safety threats.[81] A compromised control system could disable emergency stop functions, override collision detection sensors, or cause unpredictable movements that injure nearby humans. The European Union's updated Machinery Regulation explicitly recognizes this connection, requiring manufacturers to implement secure

software update mechanisms, protected control paths that prevent unauthorized commands, and documented cyber threat mitigation strategies.[81]

These requirements acknowledge that digital security and physical safety are inseparable in robots that operate autonomously among vulnerable populations.

The international standards community is working to address these gaps through initiatives like ISO 25785-1, a safety standard currently under development specifically for mobile manipulators and humanoid robots with actively controlled stability.[82] [83] This represents the first dedicated international framework designed for humanoid systems, with leadership from experts at Boston Dynamics, Agility Robotics, and other industry pioneers.[82] However, standards development proceeds through deliberate consensus processes that can take years to complete, while companies announce deployment timelines measured in months.

This tension between innovation speed and safety thoroughness creates a critical challenge. Responsible manufacturers are implementing best practices drawn from existing industrial robot standards, targeting minimum safety architecture levels that mandate redundant safety channels, diagnostic coverage to identify faults before they cause hazards, and regular validation of safety functions.[84] Yet without humanoid-specific standards, companies must interpret and apply requirements designed for different contexts, leaving gaps that could prove dangerous.[83]

The stakes are particularly high for vulnerable populations who stand to benefit most from humanoid assistance. Elderly individuals who might gain independence through robotic support, children in educational settings, patients in rehabilitation facilities, and workers in collaborative environments all deserve clear assurance that the robots entering their lives meet rigorous, verified safety standards. The robot technology's potential to improve lives makes the safety imperative more urgent, not less. Building comprehensive safety frameworks now, before widespread deployment, represents the responsible path forward and one that protects both human wellbeing and public trust in beneficial technology.

Regulation Without Stifling Innovation: What Frameworks, Policies, and Oversight Structures Should Govern Humanoid Robots

The challenge of regulating humanoid robots resembles the dilemma faced by communities deciding whether to install traffic lights at a busy intersection. Act too soon, and the infrastructure may prove unnecessary or poorly designed for actual traffic patterns. Wait too long, and preventable accidents accumulate while officials debate the details. The difference is that humanoid robots present far more complex risks than vehicle collisions, and the consequences of regulatory missteps—either through excessive restriction or insufficient oversight—will shape society for generations.

The European Union has positioned itself at the forefront of humanoid robot governance through the EU AI Act, which took effect in 2025, and the EU Machinery Regulation, which becomes enforceable in 2027.[85] These frameworks establish certifiable deployment pathways for humanoid systems in regulated sectors, providing manufacturers with clear standards they must meet before bringing products to market.[85] This approach offers predictability that enables responsible innovation while establishing baseline safety requirements that protect vulnerable populations. The EU model demonstrates that comprehensive regulation need not paralyze technological progress and can instead provide the clarity that allows beneficial applications to move forward with public confidence.[85]

China has pursued a different but equally deliberate path, establishing a national humanoid robot standards committee in November 2025 that includes representatives from major technology companies including Unitree, Xiaomi, Huawei, and XPeng.[87] The Beijing Humanoid Robot Innovation Center has led development of China's first national standards covering environmental perception, decision-making and planning, motion control, and task execution.[87][88] This rapid standardization effort reflects strategic recognition that clear frameworks accelerate rather than impede deployment by reducing uncertainty for manufacturers and users alike.

The United States has not yet established comprehensive federal regulatory frameworks equivalent to those emerging in Europe and China, creating both opportunity and risk.[85] Regulatory flexibility may enable faster innovation cycles and allow market forces to identify optimal approaches, but it also risks creating safety gaps and market fragmentation that could undermine public trust. The absence of clear standards leaves manufacturers uncertain about what requirements they will eventually face, potentially discouraging investment in safety features that exceed minimum legal obligations.

The International Organization for Standardization is developing ISO 25785-1, which will define humanoid-specific requirements including fall mitigation, predictable behavior, and compliant interactions.[85] This represents critical progress beyond existing standards like ISO 10218 and ISO/TS 15066, which focus on robot arms and collaborative robots in controlled industrial settings

rather than autonomous humanoids moving through unstructured environments.[85] Until these new standards are finalized and adopted into formal regulations, ambiguity will constrain mainstream humanoid deployment and create uncertainty for developers, investors, and potential users.[85]

The IEEE Humanoid Study Group has published a comprehensive framework developed by over sixty individuals including industry leaders, academic researchers, and regulators.[86] This framework addresses three interconnected areas: classification systems that establish clear taxonomy for different types of humanoid robots, stability standards that quantify balance and fall-response requirements, and human-robot interaction guidelines that ensure safe collaborative task performance.[86] According to Aaron Prather, director of the robotics and autonomous systems program at ASTM International and leader of the working group, the standards development work will require eighteen to thirty-six months for completion and ratification, meaning volume deployment of humanoid robots will not occur until 2027 at the earliest.[86]

This timeline provides a critical window for establishing governance structures that balance innovation with protection. Effective regulation should employ proportional oversight based on application context—high-risk deployments in healthcare, elderly care, and public safety require more stringent standards than controlled manufacturing environments. Certification pathways should specify objective performance criteria and testing protocols, reducing regulatory uncertainty while ensuring safety standards are met.[85] Perhaps most importantly, frameworks must incorporate mechanisms for periodic review and updating, ensuring that standards evolve alongside technology rather than becoming obsolete as capabilities advance.[86]

Personal and Community Preparation: Practical Steps Individuals, Families, Schools, and Towns Can Take Today to Ready Themselves for Tomorrow's Robot Integration
Waiting for regulatory frameworks to finalize before taking action could leave individuals, families, and communities unprepared for a transformation that is already underway. While standards bodies deliberate and policymakers debate, humanoid robots are moving from research laboratories into warehouses, hospitals, and pilot programs that will soon expand into broader deployment. The question facing ordinary people is not whether to prepare, but how to begin readying themselves for integration that may arrive sooner than expected.

The most fundamental preparation step involves education—not technical training in robotics engineering, but practical literacy about what humanoid robots actually are, what they can and cannot do, and how to interact with them safely. This current knowledge gap presents real risks. A person who has never encountered a humanoid robot may not understand that these machines require space to maintain balance, that sudden movements near them could trigger unexpected responses, or that their behavior depends heavily on the specific application context rather than the robot's appearance alone. Families can begin this educational process now by seeking out accurate information from reputable sources, attending public demonstrations when available, and discussing both the opportunities and limitations of humanoid technology in age-appropriate ways with children.

Schools and educational institutions occupy a particularly critical position in preparation efforts. Integrating robot safety and ethics into existing curricula long before humanoid robots become classroom fixtures builds a foundational understanding that students will carry into adulthood. Ideally, this education should extend beyond celebrating technological innovation to include honest examination of safety failures in robotics history, the importance of regulatory standards, and the shared responsibility for safe human-robot interaction. Students who understand why standards matter and how safety emerges from the interaction between humans, robots, tasks, and environments will become more thoughtful consumers, workers, and citizens in a robot-integrated world.

Communities and towns can take concrete preparatory steps even without finalized regulatory frameworks. Establishing local advisory committees to monitor humanoid robot development creates channels for information flow between technology deployers and residents. These committees can develop municipal guidelines for where humanoid robots should operate in public spaces, ensure adequate emergency response training for first responders who may encounter malfunctioning robots, and require transparency from companies planning deployments regarding their safety testing and certification status. Small towns and rural communities who are often overlooked in technology planning, particularly need these proactive structures to ensure their voices also help shape integration rather than simply having to accept whatever urban-focused companies decide to deploy.

Workplaces preparing for humanoid integration must move beyond viewing robots as simple productivity tools and recognize them as complex systems

requiring comprehensive safety management. This means conducting thorough risk assessments specific to the facility before any robot arrives, documenting all potential human-robot interaction scenarios, and identifying which workers will need training. Critically, organizations should foster safety cultures where employees feel empowered to stop robot operations if they observe unsafe conditions, and where incident reporting systems capture not just injuries but near-misses and unexpected behaviors that could inform safety improvements.

The regulatory landscape remains in flux. ISO 25785-1 and other humanoid-specific standards continue development as of 2025[85][92] with industry experts estimating that volume deployment will not occur until 2027 at the earliest once standards are finalized and ratified. This timeline provides a critical window for preparation. Individuals, families, schools, and communities that invest now in education, safety planning, and institutional readiness will be far better positioned to integrate humanoid robots safely when standards crystallize and deployment accelerates.

Perhaps most importantly, preparation requires recognizing that safety is not a fixed state achieved by purchasing certified equipment, but an ongoing process of assessment, training, and adaptation[91]. The shift in regulatory thinking from categorizing robots as inherently "collaborative" to evaluating specific "collaborative applications"[89][90] reflects a deeper truth: safety emerges from how humans and robots interact within particular contexts, not from the machines themselves[89]. Communities that internalize this principle that safety is a shared responsibility requiring vigilance, communication, and continuous learning will navigate the humanoid revolution with wisdom rather than simply reacting to whatever technology companies deliver.

The humanoid robots that will soon walk through our homes, workplaces, and communities are not arriving in some distant, imagined future—they are being tested, refined, and prepared for deployment right now. The question is no longer whether these machines will become part of daily life, but whether we will have built the safety standards, regulatory frameworks, and community readiness necessary to integrate them responsibly. The narrow window we have to get this right is closing, and the choices made in these critical years will determine whether humanoid robots become tools that genuinely serve human flourishing or technologies that erode the safety, dignity, and equity we claim to value.

The safety imperative cannot be postponed until after widespread deployment. Physical risks from collision impacts, battery system failures, and unpredictable movements in unstructured environments demand immediate attention through standardized testing protocols and clear performance benchmarks. Cybersecurity vulnerabilities that could transform helpful assistants into surveillance devices or weapons require robust protections built into design from the beginning, not patched in after breaches occur. The psychological safety of children forming attachments, elderly individuals depending on robotic care, and workers navigating hybrid human-robot environments deserves the same rigorous consideration given to physical hazards. Waiting until someone gets hurt to establish these protections would represent a failure of foresight and responsibility.

Regulatory frameworks emerging in Europe, China, and through international standards bodies demonstrate that comprehensive oversight need not stifle innovation but can instead provide the clarity and predictability that enables responsible development to move forward with public confidence. The EU AI Act and Machinery Regulation, China's national humanoid robot standards committee, and the ISO 25785-1 development process all reflect recognition that clear rules accelerate beneficial deployment by reducing uncertainty for manufacturers and users alike. The challenge lies in crafting proportional oversight that distinguishes high-risk applications requiring stringent standards from lower-risk contexts, establishing objective certification pathways, and building mechanisms for periodic review that keep pace with technological evolution.

Yet preparation cannot rest solely with policymakers and standards bodies. Individuals, families, schools, and communities have critical roles to play in readying themselves for integration that may arrive before regulatory frameworks fully crystallize. Educational efforts that build practical robot literacy, workplace safety cultures that empower employees to stop unsafe operations, municipal advisory committees that ensure local voices shape deployment decisions, and emergency response training for first responders all represent concrete steps that can begin immediately. Rural communities and small towns particularly need proactive preparation structures to ensure they are not simply receiving whatever urban-focused companies decide to deploy without consideration of their unique contexts and needs.

The most important recognition is that safety emerges not from the robots themselves, but from how humans and machines interact within specific contexts. This understanding shifts responsibility from manufacturers alone to a shared obligation involving deployers, users, regulators, and communities. The humanoid revolution will unfold safely only if everyone involved accepts their role in creating the conditions for responsible integration—from engineers designing collision detection systems to parents teaching children how to behave around robots to town councils establishing public space guidelines.

The preparation work happening now, while standards develop and deployment timelines extend into 2027 and beyond, will determine whether humanoid robots arrive as welcome assistants that enhance human capability and dignity, or as sources of injury, exploitation, and deepened inequality. The technology is coming. The only question is whether we will be ready—not just with certified equipment and finalized regulations, but with the wisdom, vigilance, and commitment to human flourishing that ensures these powerful machines truly serve the people they were built to help.

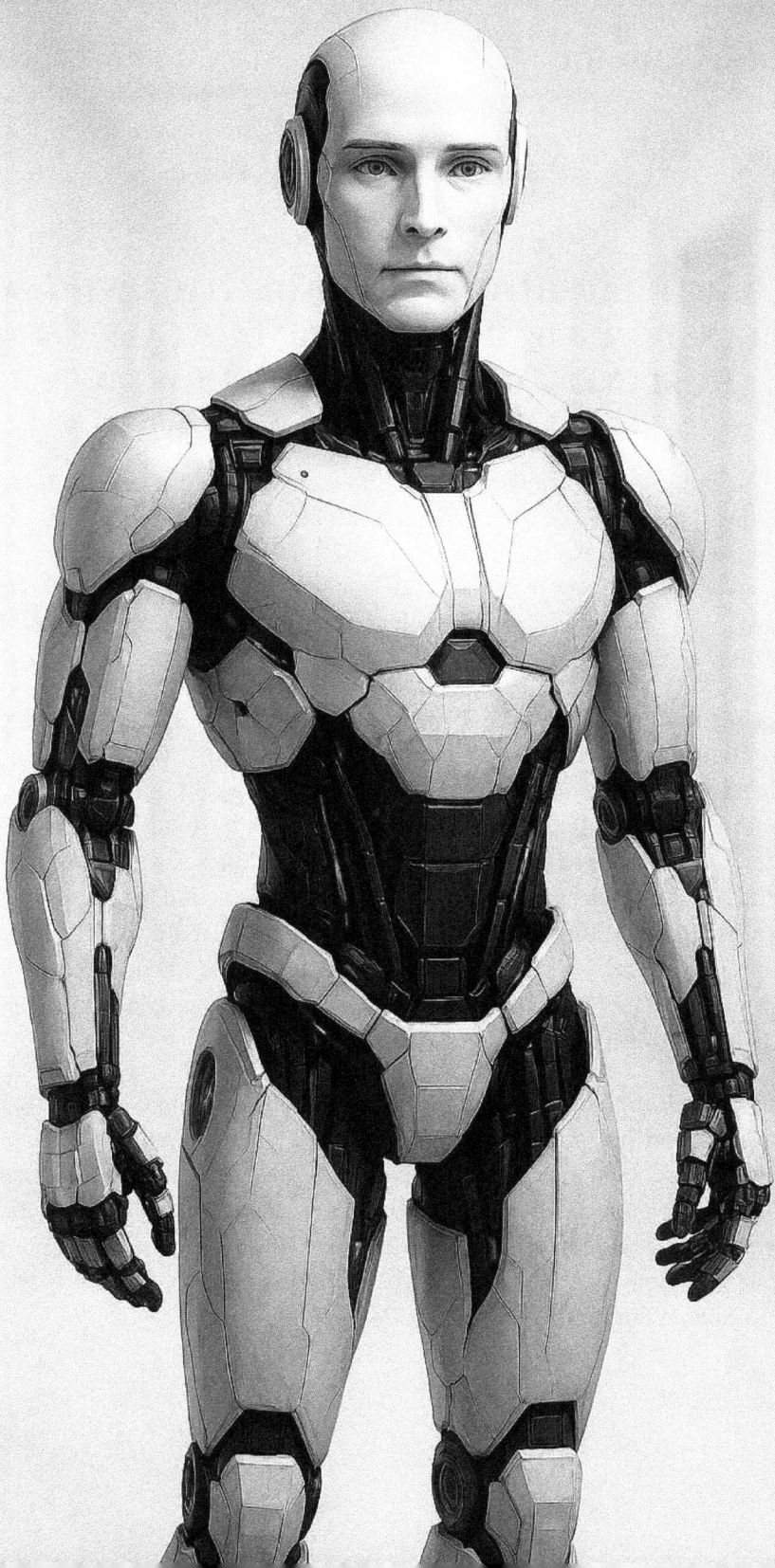

CHAPTER 10. HOW AI AND ROBOTICS WILL TRANSFORM DAILY LIVING: MAINTAINING HUMANITY IN A HUMANOID WORLD

The question isn't whether humanoid robots will change how we live as that transformation is already underway in warehouses, hospitals, and research labs around the world. The question that should keep us awake at night, the one that matters far more than technical specifications or market projections, is whether we'll allow these machines to change who we are.

Throughout this book, we've explored the practical realities of humanoid robots entering our workplaces, homes, and communities. We've examined job displacement and economic adaptation, elder care possibilities and ethical boundaries, childhood development concerns and rural infrastructure challenges. We've looked honestly at what these machines can and cannot do, at the timeline for their arrival, and at the safety and regulatory frameworks we desperately need. But now, as we approach the threshold of widespread humanoid integration, we must confront the most fundamental question of all: How do we maintain our essential humanity when machines can increasingly replicate human form, function, and interaction?

This isn't a philosophical abstraction. It's a practical challenge that families, communities, and individuals will face daily—the choice between asking a robot or calling a neighbor for help, between more efficient automated care and slower human assistance, between the convenience of machine interaction and the messiness of genuine human connection. Each choice seems minor in isolation, but together they will shape whether humanoid robots amplify what makes us human or gradually erode it.

The paradox we face is real and uncomfortable. Humanoid robots promise to free us from drudgery, extend our capabilities, and solve pressing problems ranging from labor shortages to elder care crises. Yet the very efficiency and convenience they offer could inadvertently diminish the experiences, relationships, and struggles that forge meaning, character, and connection for humans. When robots can perform caregiving tasks, will we lose the patience and presence that caregiving teaches? When machines handle household labor, will we miss the shared rhythms and small collaborations around chores that bind families together? When automation replaces difficult work, will we lose the sense of purpose and identity that meaningful labor provides?

These aren't reasons to reject humanoid robots—they're reasons to approach their integration with intentionality. The technology itself is neither good nor evil; it's a powerful tool that will reflect the values, priorities, and choices we bring to its design and deployment. The future we're building isn't predetermined by engineering capabilities or market forces. It will be shaped by the boundaries we set, the uses we permit, the regulations we demand, and most importantly, by the daily choices we make about when to embrace robotic assistance and when to insist on irreplaceable human presence.

This chapter explores how we navigate that balance—not through abstract principles, but through practical frameworks for preserving human connection, cultivating irreplaceable human capacities, and designing our robot-integrated lives to serve genuine flourishing rather than mere convenience. The humanoid age doesn't have to diminish our humanity. If we're thoughtful, intentional, and brave enough to prioritize what truly matters, it might actually help us rediscover it.

The Humanity Paradox: What We Risk Losing (and What We Might Gain) When Robots Handle Life's Daily Tasks

The promise sounds almost utopian: robots that fold laundry, prepare meals, remind elderly parents to take medications, and handle the endless stream of household tasks that consume hours each day. Time, that most precious and finite resource, suddenly becomes abundant. Families could spend evenings together instead of rushing through chores. Caregivers could focus on emotional connection rather than physical labor. Workers could pursue creative endeavors instead of repetitive tasks. This is the gain side of the humanity paradox—genuine liberation from the drudgery that has defined human existence for millennia.

Yet something more complex unfolds when machines handle life's daily tasks. Research on social robots in elder care reveals this tension clearly. Older adults using companion robots showed participants averaging 126 interactions over two weeks, describing the machines as "private secretaries" and even "members of the family."[93] The robots provided calendar reminders, weather updates, music, and conversation—practical assistance that addressed real isolation and need.[93] But the research also revealed a subtle shift: participants used robots as substitutes for human interaction, not supplements to it.[93] The convenience was real, but so was the displacement of human connection.

This paradox extends far beyond elder care into the texture of ordinary family life. Consider the parent who derives meaning from cooking meals for their children—not just the nutrition provided, but the act itself, the passing down of recipes, the conversations that happen while chopping vegetables together. When a humanoid robot handles meal preparation with greater efficiency and consistency, the family gains time but potentially loses the small rituals through which relationships are built and maintained. The task becomes optional rather than necessary, and with that shift, something intangible but essential may fade.

The same pattern appears in caregiving contexts. When adult children care for aging parents, physical tasks such as helping with bathing, managing medications, and preparing meals are undeniably burdensome. Yet these acts of care also carry moral weight and relational significance. The human act of thinking about another person's wellbeing, of holding them in mind throughout the day, creates bonds that transcend the tasks themselves. A robot can deliver medication reminders more reliably than any human, but the automation erases the relational dimension entirely. The parent knows someone is thinking of them not because an algorithm triggered a notification, but because their child chose to remember.

The broader social Implications compound these Individual losses. Humans have historically derived purpose from work, from the necessity of caring for one another, from managing the basic demands of existence. These activities build competence, resilience, and connection to both the material world and other people. When robots assume these roles, particularly for children and young adults who have never known life without such assistance, society risks raising generations disconnected from the formative experiences that develop character and capability.

This is not an argument for unnecessary suffering or the rejection of helpful technology. The elderly person living alone who uses a robot for medication reminders and companionship experiences genuine benefit, especially when human support is unavailable.[93] The exhausted caregiver who receives robotic assistance with physically demanding tasks gains capacity to provide the emotional presence that machines cannot replicate. The question is not whether to use humanoid robots, but how to integrate them without surrendering the experiences, relationships, and struggles that make us fully human.

The humanity paradox demands intentionality. Society must consciously preserve spaces where human labor and care remain central, even when robots could perform tasks more efficiently. Families must deliberately choose which tasks to automate and which to protect as fundamentally human domains. Communities must ensure that technological convenience serves human flourishing rather than merely optimizing away the messy, inefficient, irreplaceable experiences through which meaning emerges. The gains are real, but so are the risks, and navigating between them will define whether the humanoid age enhances or diminishes what it means to be human.

Intentional Living in a Robot World: Setting Boundaries, Protecting Connection, and Designing Technology Integration That Serves Human Flourishing

The difference between technology that serves human flourishing and technology that erodes it often comes down to a single word: intentionality. Without conscious boundaries and deliberate design choices, humanoid robots will integrate into daily life according to market forces, convenience, and the path of least resistance—not according to what genuinely serves human wellbeing, connection, and meaning. The question families, communities, and individuals must answer is not whether to live alongside humanoid robots, but how to do so in ways that amplify rather than diminish what makes life worth living.

Setting boundaries in a robot world begins with distinguishing between genuine need and mere convenience. When humanoid robots address labor shortages in hazardous industrial environments, they serve a clear human need—protecting workers from injury while maintaining productivity.[21] When they provide physical assistance to elderly individuals who would otherwise lose independence, they enable dignity and autonomy that would be impossible without technological support.[21] But when robots replace human interaction simply because automation is more efficient, or when they handle tasks that build competence and connection in children simply because it's easier for them than the parents, the calculus changes entirely. The boundary worth protecting is this: humanoid robots should augment human capability and address genuine deficits, not optimize away the experiences through which meaning, relationship, and character develop.

Protecting connection in a humanoid world requires conscious choices about which domains remain fundamentally human. Certain activities carry relational weight beyond their practical function—the parent who cooks meals with children, the adult child who manages an aging parent's medications, the neighbor who checks in during difficult times. These acts create bonds through the simple fact of holding another person in mind and choosing to remember their wellbeing.

Families navigating this landscape might establish clear guidelines: robots handle physically demanding tasks that cause caregiver burnout, but humans maintain responsibility for emotional presence and decision-making. Communities might deploy humanoid robots for hazardous public maintenance while preserving human roles in education, counseling, and social services where relationship is therapeutic.[21] Workplaces might use robots for ergonomic relief and safety while preserving human collaboration, problem-solving, and the sense of purpose that meaningful work provides.[21]

The three-horizon deployment timeline for humanoid robots—from controlled industrial settings to semi-structured service environments to open-ended real-world contexts—offers a framework for gradual, intentional integration.[94] [8] This phased approach allows society to learn what works, what preserves human flourishing, and where boundaries matter most before humanoid robots become ubiquitous. The goal is not to resist helpful technology, but to ensure that each expansion of robot capability serves genuine human needs rather than merely optimizing for efficiency at the expense of connection, meaning, and the irreplaceable experiences that make us fully human.

The Irreplaceable Human: Cultivating the Capacities, Relationships, and Experiences That Will Matter Most in an Automated Age
As humanoid robots assume more tasks that once required human hands and minds, a crucial question emerges: what remains uniquely, irreplaceably human? The answer matters profoundly, because the capacities, relationships, and experiences worth protecting and cultivating will determine whether the humanoid age enhances human flourishing or gradually erodes it. Understanding what cannot be automated and choosing to preserve those domains even when automation becomes possible represents perhaps the most important decision individuals and communities will make in the coming decades.

The capacities that remain distinctly human are not those involving speed, consistency, or computational power (domains where machines already excel). Instead, they are the qualities that emerge from embodied, conscious existence: the ability to find meaning in struggle, to form relationships characterized by genuine vulnerability and mutual growth, to exercise moral judgment in ambiguous situations where rules provide insufficient guidance, and to create beauty and purpose from the raw materials of finite existence. These capacities cannot be programmed or replicated by algorithms, no matter how sophisticated humanoid robots become.

Consider the difference between a robot that reminds an elderly person to take medication and an adult child who calls daily to check in. Both accomplish the practical task of medication management, but only the human interaction carries relational weight. The parent knows someone is thinking about them and choosing to remember their wellbeing throughout the day. This distinction between automated function and chosen presence defines much of what makes human connection irreplaceable. The capacity to be genuinely affected by another person's suffering or joy, to be transformed by relationships rather than merely executing programmed responses, creates bonds that technology cannot simulate.

Moral agency represents another irreplaceable human dimension. While humanoid robots can be programmed with ethical guidelines and decision trees, they cannot exercise the kind of moral deliberation that emerges from lived experience, reflection on consequences, and engagement with diverse perspectives. The responsibility for consequential decisions—particularly those affecting vulnerable populations like children, elderly individuals, or communities facing difficult tradeoffs—must remain with humans who can be held accountable by other moral agents. Delegating these decisions to machines, even highly capable ones, represents an abdication of human responsibility that threatens the foundations of ethical community.

Creativity and meaning-making likewise remain distinctly human. Humanoid robots can generate novel combinations of existing patterns, but human creativity involves envisioning genuinely new possibilities and investing those imaginings with personal significance. A robot might compose music by analyzing patterns in existing compositions, but it cannot experience the emotional resonance that makes music meaningful to human listeners. The human capacity to construct narratives that transform experience into significance—to find purpose in adversity, to create beauty that speaks to shared human experience—emerges from consciousness and embodied existence that machines do not possess.

Cultivating these irreplaceable capacities requires intentional choices about education, community structure, and technology integration. Educational systems must shift from emphasizing information transfer (where machines excel) toward developing critical thinking, ethical reasoning, emotional intelligence, and the capacity for authentic relationships. Communities might deliberately preserve domains where human presence matters most: caregiving contexts where relationship is therapeutic, creative endeavors where human vision remains central, civic participation where collective deliberation shapes shared futures, and skilled trades where embodied expertise and aesthetic judgment prove irreplaceable.

The

deepest human irreplaceability lies in embodied, finite existence itself. Humans inhabit physical bodies that experience pleasure and pain, hunger and satisfaction, fatigue and vitality. This embodiment grounds human experience in ways that disembodied information processing cannot replicate. Furthermore, awareness of mortality—the knowledge that existence is finite—

profoundly shapes human meaning-making. The urgency created by finitude motivates the deepening of relationships, the creation of legacies, and the pursuit of purposes that transcend individual survival. These dimensions emerge fundamentally from the human condition itself, and they define what will matter most as humanoid robots increasingly handle the practical tasks of daily existence.[95]The humanoid age will arrive not with a single dramatic moment, but through thousands of small decisions like choices made by families about which tasks to automate, communities choosing where robots serve and where humans remain essential, and individuals deciding what capacities to cultivate and what relationships to protect. These decisions will collectively determine whether humanoid robots enhance human flourishing or gradually erode the experiences, connections, and struggles that make life meaningful.

The paradox at the heart of this transformation remains unresolved and perhaps unresolvable: the same technology that promises liberation from drudgery also threatens to optimize away the formative experiences through which character, competence, and connection develop. Robots that handle caregiving tasks free exhausted family members to provide emotional presence—yet they also risk replacing the acts of care through which bonds deepen and moral responsibility finds expression. Machines that manage household labor create time for creativity and relationship—yet they also remove the shared rhythms and small collaborations around chores that bind families together across generations. This tension cannot be eliminated through better engineering or smarter algorithms. It can only be navigated through intentionality, wisdom, and the courage to prioritize what truly matters over what is merely convenient.

The capacities worth protecting an" cul'Ivating in an automated age are precisely those that emerge from embodied, conscious, finite human existence. Genuine vulnerability in relationships, moral deliberation in ambiguous situations, creativity that transforms experience into meaning, and the urgency that mortality brings to human purpose—these dimensions cannot be programmed or replicated, no matter how sophisticated humanoid robots become. Educational systems, community structures, and family practices must deliberately preserve domains where these irreplaceable human qualities remain central, even when automation offers apparent efficiency gains.

The future being built is not predetermined by engineering capabilities or market forces. It will reflect the values, priorities, and choices that individuals, families, and communities bring to humanoid robot integration. Technology itself remains morally neutral as a powerful tool that will serve whatever purposes humans assign it. The question is not whether humanoid robots will change daily life, but whether that change will be guided by genuine concern for human wellbeing or by the path of least resistance toward efficiency and convenience.

This moment demands both hope and vigilance. Hope, because humanoid robots genuinely can address pressing challenges including labor shortages in hazardous industries, elder care crises, accessibility barriers for people with disabilities, and the physical burdens that prevent aging adults from maintaining independence and dignity. Vigilance, because the same technology can displace meaningful work, replace irreplaceable human connection, and gradually erode the capacities that define human flourishing. The difference between these outcomes lies entirely in the intentionality humans bring to design, deployment, and daily choices about when to embrace robotic assistance and when to insist on human presence.

The humanoid age does not have to "lmin'sh humanity. If society approaches this transformation with wisdom, compassion, and an unwavering commitment to what makes life worth living, it might actually help rediscover what matters most—not by providing answers through automation, but by forcing conscious choices about the relationships, experiences, and values worth protecting in a world where nearly everything can be delegated to machines.

CONCLUSION

The elderly woman in the rehabilitation facility reaches for the humanoid robot's hand, and the machine responds with careful precision—the right pressure, the right timing, the gesture programmed to comfort. Watching this interaction, we might ask ourselves: Is this the future we want? But perhaps that's the wrong question. The right question is: How do we shape this future so it serves our deepest human needs rather than diminishing them?

Throughout this book, we've explored the landscape of a world where humanoid robots walk among us—not as distant science fiction, but as an approaching reality that will reshape our homes, workplaces, communities, and relationships. We've examined what these machines can actually do, distinguished hype from genuine capability, and confronted both the extraordinary opportunities and the serious risks they present. We've looked honestly at job displacement and economic disruption, at the promise of dignified elder care and the danger of replacing human connection with technological substitutes, at children growing up with robot companions and rural communities facing infrastructure barriers that could deepen existing inequalities.

The humanoid revolution is not something happening to us—it's something we're creating, and we still have agency in how it unfolds. The robots being designed today in research labs and corporate facilities will reflect the values, priorities, and ethical boundaries we establish now, not later. If we want humanoid robots that enhance human dignity rather than compromise it, that reduce inequality rather than amplify it, that support genuine human connection rather than replace it, we must participate actively in shaping their development, deployment, and integration into our lives.

This means demanding robust safety standards before humanoid robots enter our homes and communities, not after accidents reveal the gaps in oversight.

It means insisting on regulation that protects workers, consumers, and vulnerable populations while still allowing beneficial innovation to proceed. It means having difficult conversations in our families about what roles we want robots to play and what boundaries we need to protect. It means ensuring that rural communities, small towns, and underserved populations aren't left behind as robot benefits flow primarily to wealthy urban areas.

Most importantly, it means cultivating and protecting the irreplaceable human capacities that will matter most in an automated age. Humanoid robots may eventually handle countless practical tasks, but they cannot replicate the genuine warmth of human presence, the creativity born from struggle and spontaneity, the moral wisdom that comes from lived experience, or the authentic relationships that give life meaning. As robots take on more of life's functional demands, we must become more intentional about preserving space for what makes us human—the messy, inefficient, beautiful aspects of life that resist automation.

The future we're entering is neither utopian nor dystopian—it's simply human, with all the complexity, contradiction, and possibility that entails. Humanoid robots will help some people maintain independence and dignity while potentially isolating others. They'll eliminate dangerous jobs while displacing some workers who will therefore need new pathways forward. They'll provide care in places that desperately need it while raising profound questions about what care truly means.

We stand at a threshold watching as machines that look like us prepare to step into our world. What happens next depends not on the robots themselves, but on the choices we make about how to receive them. If we approach this transformation with wisdom, compassion, and commitment to human flourishing and most importantly remember that technology should serve life rather than the other way around, we can build a future where humanoid robots walk among us, not as replacements for humanity, but as tools that help us become more fully human. That future is still ours to create.

YOUR VOICE TRULY MATTERS

If the book helped you learn something new, feel more empowered about your health, or simply sparked your curiosity, sharing that experience makes a meaningful difference. Simply scan the QR code.

Thank you so much for your time, your support, and for being part of this artificial intelligence journey.

With Gratitude,
Alexandria Isaacs.

BIBLIOGRAPHY

REFERENCES

[1] Isaac, S. *What is a Humanoid Robot: Beginner's Guide*. Qviro. **https://qviro.com/blog/what-is-a-humanoid-robot/**

[2] Meegle. (2025, January 10). *Humanoid Robots*. Meegle. **https://www.meegle.com/en_us/topics/robotics/humanoid-robots**

[3] Fankhauser, D. (2025, March 11). *What Is a Humanoid? Definition and Examples*. Robozaps. **https://blog.robozaps.com/b/what-is-a-humanoid**

[4] NVIDIA. *What Is a Humanoid Robot?* NVIDIA. **https://www.nvidia.com/en-us/glossary/humanoid-robot/**

[5] Fiveable. *Humanoid Robots*. Fiveable. **https://fiveable.me/key-terms/ap-japanese/humanoid-robots**

[6] Humanoid Robotics Technology. (2025, February). *Humanoid: A Complete Guide to Humanoid Robots*. Humanoid Robotics Technology. **https://humanoidroboticstechnology.com/articles/humanoid-complete-guide-humanoid-robots/**

[7] Standard Bots. (2025, August 7). *The most advanced robots in 2025*. Standard Bots. **https://standardbots.com/blog/most-advanced-robot**

[8] Bain & Company. (2025). *Humanoid Robots: From Demos to Deployment*. Bain & Company. **https://www.bain.com/insights/humanoid-robots-from-demos-to-deployment-technology-report-2025/**

[9] Simmie, S. (2024, October 8). *Dual manipulator Rosie the robot used for Industry 4.0 research*. InDro Robotics. **https://indrorobotics.ca/dual-manipulator-rosie-the-robot-used-for-industry-4-0-research/**

[10] Novak, M. (2012, November 11). *Recapping the Jetsons Episode 08 Roseys Boyfriend.* Smithsonian Magazine. **https://www.smithsonianmag.com/history/recapping-the-jetsons-episode-08-roseys-boyfriend-120640948/**

[11] Yearsdon, A. (2025, April 28). *Humanoid Robots Guide (2025): Types, History, Best Models, Anatomy and Applications.* Top3DShop. **https://top3dshop.com/blog/humanoid-robots-types-history-best-models**

[12] Kalil, M. *Humanoid Robots Timeline: Ancient Automata to 2075.* Mike Kalil. **https://mikekalil.com/blog/humanoid-robot-timeline/**

[13] Thompson, C. *13 Milestones in the History of Robotics.* Aventine. **https://aventine.org/robotics/history-of-robotics**

[14] Standard Bots. (2025, August 8). *Top AI robotics companies to watch in 2025 (and what they're actually building).* Standard Bots. **https://standardbots.com/blog/ai-robotics-companies**

[15] Isaac, S. (2025). *Humanoid Robot Landscape 2025.* Qviro. **https://qviro.com/blog/humanoid-robot-landscape-2025/**

[16] Mapue, J. (2025, July 14). *21 top companies in the vanguard of the rise of humanoid robots.* Ross Dawson. **https://rossdawson.com/futurist/companies-creating-future/top-companies-rise-humanoid-robots/**

[17] Varanasi, L. (2025, November 27). *6 leading humanoid robot companies worth watching.* Business Insider. **https://www.businessinsider.com/humanoid-robot-companies-us-tesla-figure-1x-agility-apptronik-2025-11**

[18] Fankhauser, D. (2025, April 8). *The Economic Impact of Humanoid Robots on the Job Market.* Robozaps. **https://blog.robozaps.com/b/economic-impact-of-humanoid-robots-on-job-market**

[19] Brown, S. *A new study measures the actual impact of robots on jobs. It's significant.* MIT Sloan. **https://mitsloan.mit.edu/ideas-made-to-matter/a-new-study-measures-actual-impact-robots-jobs-its-significant**

[20] Smith, T. (2025, March 6). *The uneven labor market impact of industrial robots.* American Economic Association. **https://www.aeaweb.org/research/automation-employment-gaps-us**

[21] Neumann, C. (2025, June). *Humanoid robots offer disruption and promise.* World Economic Forum. **https://www.weforum.org/stories/2025/06/humanoid-robots-offer-disruption-and-promise/**

[22] Fankhauser, D. (2025, February 6). *Humanoid Robots in the Workplace: Transforming Efficiency and Productivity.* Robozaps. **https://blog.robozaps.com/b/humanoid-robots-in-workplace**

[23] Capitol Technology University. (2024, January 22). *From Science Fiction to Reality: The Rise of Humanoid Robots in the Workplace.* Capitol Technology University. **https://www.captechu.edu/blog/science-fiction-reality-rise-of-humanoid-robots-workplace**

[24] Cogent Infotech. *Robots As Colleagues: How Collaborative Robots (Cobots) Are Changing The Workforce.* Cogent Infotech. **https://www.cogentinfo.com/resources/robots-as-colleagues-how-collaborative-robots-cobots-are-changing-the-workforce**

[25] Humanoid. (2025, July 22). *The Future of Work: How Humanoid Robots Will Revolutionize Industries.* RoboticsTomorrow. **https://www.roboticstomorrow.com/article/2025/06/the-future-of-work-how-humanoid-robots-will-revolutionize-industries/24932**

[26] Bain & Company. (2025, April). *Humanoid robots at work: What executives need to know.* Bain & Company. **https://www.bain.com/insights/humanoid-robots-at-work-what-executives-need-to-know/**

[27] Fankhauser, D. (2025, April 8). *The Best Humanoid Robots Available in 2025.* Robozaps. **https://blog.robozaps.com/b/best-humanoid-robots**

[28] Contreras, C. (2025, October 02). *Humanoid robots at home: Google DeepMind showcases AI-powered household helpers.* Northeastern Global

News. https://news.northeastern.edu/2025/10/02/humanoid-robots-at-home-deepmind-google/

[29] Kuhns, E. J. (2025, January 15). *BEST AI PERSONAL ROBOTS I SAW AT CES 2025!* Eric J. Kuhns. https://www.ericjkuhns.com/blog/best-tech-i-saw-from-ces-2025-robots-ai-and-more

[30] Fankhauser, D. (2025, April 8). *Humanoid Robots in Elderly Care: Enhancing Quality of Life.* Robozaps. https://blog.robozaps.com/b/humanoid-robots-in-elderly-care

[31] Chu, J. (May 13, 2025). *Eldercare robot helps people sit and stand, and catches them if they fall.* MIT News. https://news.mit.edu/2025/eldercare-robot-helps-people-sit-stand-catches-them-fall-0513

[32] *Age In Place With Robots.* Live In Home Care. https://www.liveinhomecare.com/age-in-place-with-robots/aging-in-place/

[33] Blue Frog Robotics. *Buddy: An Innovative Solution for Assisting Seniors at Home and in Care Facilities.* Blue Frog Robotics. https://www.bluefrogrobotics.com/buddy-an-innovative-solution-for-assisting-seniors-at-home-and-in-care-facilities

[34] Dimov, D. (2015, October 16). *Privacy Risks of Household Robots: 5 Security Risks and 10 Steps to Protect Yourself.* Infosec Institute. https://www.infosecinstitute.com/resources/general-security/privacy-risks-of-household-robots-5-security-risks-and-10-steps-to-protect-yourself/

[35] Lutz, C., & Tamò-Larrieux, A. (2021, April 26). *Do Privacy Concerns About Social Robots Affect Use Intentions? Evidence From an Experimental Vignette Study.* Frontiers in Robotics and AI. https://pmc.ncbi.nlm.nih.gov/articles/PMC8110194/

[36] Afroze, D., Tu, Y., & Hei, X.. (2024). *Securing the Future: Exploring Privacy Risks and Security Questions in Robotic Systems.* arXiv. https://arxiv.org/html/2409.09972v1

[37] Relford, E. (2025, March 5). *Privacy in the age of robotics.* IAPP. **https://iapp.org/news/a/privacy-in-the-age-of-robotics**

[38] Wynn, P. (2025, July 24). *Exclusive: AARP-NAC Report Finds 45% Increase in Americans Providing Care.* AARP. **https://www.aarp.org/caregiving/basics/caregiving-in-us-survey-2025/**

[39] National Council on Aging. (2025). *Addressing the Nation's Retirement Crisis: The 80 Percent Financially Struggling.* National Council on Aging. **https://www.ncoa.org/article/addressing-the-nations-retirement-crisis-the-80-percent-financially-struggling/**

[40] Margolis, H. S. (2025, October 23). *The Coming Elder Care Challenge: More People Are Beginning To Notice.* Center for Retirement Research. **https://crr.bc.edu/the-coming-elder-care-challenge-more-people-are-beginning-to-notice/**

[41] Kuwik, A. (2025, May 1). *The State of Aging in Colorado 2025.* Bell Policy Center. **https://bellpolicy.org/the-state-of-aging-in-colorado-2025/**

[42] Johns Hopkins Bloomberg School of Public Health. (2025, July 28). *What Is the Caregiver Crisis?* Johns Hopkins Bloomberg School of Public Health. **https://publichealth.jhu.edu/2025/what-is-the-caregiver-crisis**

[43] Nestler, M. (2025, February 5). *The care crisis: Eldercare collides with childcare.* KPMG. **https://kpmg.com/us/en/articles/2025/february-2024-the-care-crisis-eldercare.html**

[44] Meegle. (2025, January 3). *Robots For Elderly Care.* Meegle. **https://www.meegle.com/en_us/topics/robotics/robots-for-elderly-care**

[45] Vanjare, S. *Robotics in Elderly Care: Innovations, Benefits, & Future Trends.* Inkwood Research. **https://www.inkwoodresearch.com/robotics-in-elderly-care/**

[46] Nature. (2024). *Are robots the solution to the crisis in older-person care?* Nature. **https://www.nature.com/articles/d41586-024-01184-4**

[47] Badawy, A. (2024, November 12). *Social robots with older people in long-term care homes: Ethical perspectives in light of caring traditions.* scup.com. **https://www.scup.com/doi/10.18261/9788215069234-24-10**

[48] Fankhauser, D. (2025, March 5). *Humanoid Robots in Education: Transforming Classrooms.* Robozaps. **https://blog.robozaps.com/b/humanoid-robots-in-education**

[49] Raptopoulou, A., Komnidis, A., Bamidis, P. D., & Astaras, A. (2021, May 19). *Human–robot interaction for social skill development in children with ASD: A literature review.* Healthcare Technology Letters. **https://pmc.ncbi.nlm.nih.gov/articles/PMC8284575/**

[50] Takata, K., Yoshikawa, Y., Muramatsu, T., Matsumoto, Y., Ishiguro, H., Mimura, M., & Kumazaki, H. (2023, July 19). *Social skills training using multiple humanoid robots for individuals with autism spectrum conditions.* Frontiers in Psychiatry. **https://pmc.ncbi.nlm.nih.gov/articles/PMC10394831/**

[51] VUMC News and Communications. (2013, March 23). *Humanoid robot helps train children with autism.* VUMC News. **https://news.vumc.org/2013/03/23/robot-helps-children-with-autism/**

[52] Baynes, R. (2024, October 22). *The Benefits of Robotics and AI for Children and Behavioral Health.* Behavioral Health News. **https://behavioralhealthnews.org/the-benefits-of-robotics-and-ai-for-children-and-behavioral-health/**

[53] İnan, B., & Güldenoğlu, B. (2025, January 3). *The Use and Future of Social Humanoid Robots in Special Education: A Systematic Review.* International Journal of Sciences: Basic and Applied Research (IJSBAR). **https://www.gssrr.org/JournalOfBasicAndApplied/article/view/17263**

[54] PredictML.ai. *Humanoid Robots in Education: How AI is Transforming Classrooms with EDUBOT.* PredictML.ai. **https://predictml.ai/humanoid-robots-in-education-transforming-classrooms-edubot/**

55 3Laws Robotics. *Humanoids and the Future of Educational robots*. 3Laws Robotics. **https://3laws.io/pages/Humanoids_and_the_Future_of_Educati onal_robots.html**

56 Becker,   H.   (2018,   October   2). *Hubots*.   Kids   Can Press. **https://www.kidscanpress.com/product/hubots/**

57 Fankhauser, D. (2025, February 13). *Humanoid Robots in Agriculture: Enhancing Efficiency*. Robozaps. **https://blog.robozaps.com/b/humanoid-robots-in-agriculture**

58 Fresh Consulting. *Robots in Agriculture: Transforming the Future of Farming*.   Fresh Consulting. **https://www.freshconsulting.com/insights/blog/robots-in-agriculture-transforming-the-future-of-farming/**

59 Standard Bots. (2025, May 5). *How agriculture robots are changing the farming   industry*.   Standard Bots. **https://standardbots.com/blog/agriculture-robots**

60 BioIntelliSense. (2025). *Rural Healthcare Transformation with Scalable Continuous   Monitoring*. BioIntelliSense. **https://www.biointellisense.com/rhtp/**

61 South Dakota Association of Healthcare Organizations. (2025, August 29). *Rural Initiatives: AI, automation & robotics in health care roles*. SDAHO. **https://sdaho.org/2025/08/29/rural-initiative-ai-automation-robotics-in-health-care-roles/**

62 Rural Health Information Hub. (2024, August 6). *Applying AI to Rural Health,   with   Jordan   Berg*.   Rural   Health   Information Hub. **https://www.ruralhealthinfo.org/podcast/ai-aug-2024**

63 Guo, J., & Li, B. (2018, August 1). *The Application of Medical Artificial Intelligence Technology in Rural Areas of Developing Countries*. Health Equity. **https://pmc.ncbi.nlm.nih.gov/articles/PMC6110188/**

64 Humanoid Robotics Technology. (2025, May). *How Humanoid Robots Are Transforming   Healthcare*.   Humanoid   Robotics

Technology. **https://humanoidroboticstechnology.com/articles/how-humanoid-robots-are-transforming-healthcare/**

[65] Schwartz (2024, September 30). *Bridging the Digital Divide: Enhancing Broadband Access in Rural Affordable Housing.* Capcon Networks. **https://www.capconnetworks.com/bridging-the-digital-divide-enhancing-broadband-access-in-rural-affordable-housing**

[66] Cooper, L. (2023, October 6). *The Stakes Are Too High to Not Solve the Rural Digital Divide.* Human-I-T. **https://www.human-i-t.org/why-bridge-rural-digital-divide/**

[67] APCO International. *Bridging the Digital Divide: Navigating the Challenges of Rural Broadband Deployment.* APCO International. **https://www.apcointl.org/psc/bridging-the-digital-divide-navigating-the-challenges-of-rural-broadband-deployment-2/**

[68] Taylor, J. (2025). *Bridging the Gap: The Digital Divide in Rural America and Its Impact on Black Women and Families.* NCNW. **https://ncnw.org/bridging-the-gap-the-digital-divide-in-rural-america-and-its-impact-on-black-women-and-families/**

[69] Turner Lee, N. *Closing the digital and economic divides in rural America.* Brookings. **https://www.brookings.edu/articles/closing-the-digital-and-economic-divides-in-rural-america/**

[70] Zhao, L. (2023, September 7). *How the Digital Divide Affects America's Rural Small Businesses.* Federal Reserve Bank of Cleveland. **https://www.clevelandfed.org/publications/notes-from-the-field/2023/nftf-20230907-how-the-digital-divide-affects-americas-rural-small-businesses**

[71] House of Ethics. (2024, August 04). *Human Rights, Robot Wrongs: Being Human in The Age of AI.* House of Ethics. **https://www.houseofethics.lu/2024/08/04/human-rights-robot-wrongs-by-susie-alegre/**

[72] PONS IP. (2022, February 25). *SMART ROBOTS: LEGAL ASPECTS AND RESPONSIBILITY.* Pons IP. **https://ponsip.com/en/ip-news/uncategorized/smart-robots-legal-aspects-and-responsibility/**

[73] Ascension Nexus® Law. (2025, April 2). *The Rise of Autonomous Robots: Legal and Ethical Considerations for the Home of the Future*. Estate and Family Lawyer. **https://estateandfamilylawyer.com/the-rise-of-autonomous-robots-legal-and-ethical-considerations-for-the-home-of-the-future/**

[74] Kirkpatrick, K. (2013, November 1). *Legal Issues with Robots.* Communications of the ACM. **https://cacm.acm.org/news/legal-issues-with-robots/**

[75] JusCorpus. (March 31, 2025). *Should Robots Have Legal Rights? The Debate on AI Personhood*. JusCorpus. **https://www.juscorpus.com/should-robots-have-legal-rights-the-debate-on-ai-personhood/**

[76] Stanford, E. (2025, May 15). *Autonomous AI: who is responsible when AI acts autonomously and things go wrong?* Global Legal Insights. **https://www.globallegalinsights.com/practice-areas/ai-machine-learning-and-big-data-laws-and-regulations/autonomous-ai-who-is-responsible-when-ai-acts-autonomously-and-things-go-wrong/**

[77] Weir, K. (2018, January 1). *Can we design moral robots?* Monitor on Psychology. **https://www.apa.org/monitor/2018/01/moral-robots**

[78] Wright, C. *Ethical concerns with replacing human relations with humanoid robots: an Ubuntu perspective*. Montreal AI Ethics Institute. **https://montrealethics.ai/ethical-concerns-with-replacing-human-relations-with-humanoid-robots-an-ubuntu-perspective/**

[79] Friedman, C. (2025, March 28). *Artefacts of Change: The Disruptive Nature of Humanoid Robots Beyond Classificatory Concerns*. PMC (PubMed Central). **https://pmc.ncbi.nlm.nih.gov/articles/PMC11953219/**

[80] Steele, A. (2025, October 26). *The Ethics of Humanoid Robots*. USC Viterbi Center for Engineering. **https://vce.usc.edu/weekly-news-profile/the-ethics-of-humanoid-robots/**

[81] Chen, B. (2025, October 28). *A Primer On Humanoid Robot Compliance: Safety, Standards, And The Path To Public Trust*. Kite Compliance. **https://www.kitecompliance.ai/vertical-compliance/humanoid-robot-compliance**

[82] Sethi, C. (2025, May 14). *Safety in Motion: Setting the Standard for Humanoid Robots.* Robotics & Automation INSIDER. **https://www.techbriefs.com/component/content/article/53111-safety-in-motion-setting-the-standard-for-humanoid-robots**

[83] Humanoids Daily. *Novanta hosts key ISO meeting to codify humanoid robot safety rules.* Humanoids Daily. **https://www.humanoidsdaily.com/feed/novanta-hosts-key-iso-meeting-to-codify-humanoid-robot-safety-rules**

[84] Texas Instruments. (2025). *Application Brief Power Stage Implementations for Humanoid Robots.* Texas Instruments. **https://www.ti.com/lit/SLVAG23**

[85] Kelkar, A., Jansen, C., Chu, F., Patel, M., Chui, M., & Robertson, M. (2025, October 15). *Humanoid robots: Crossing the chasm from concept to commercial reality.* McKinsey & Company. **https://www.mckinsey.com/industries/industrials-and-electronics/our-insights/humanoid-robots-crossing-the-chasm-from-concept-to-commercial-reality**

[86] Oitzman, M. *IEEE study group publishes framework for humanoid standards.* The Robot Report. **https://www.therobotreport.com/ieee-study-group-publishes-framework-for-humanoid-standards/**

[87] TechNode. (2025, November 25). *China unveils humanoid robot standards committee with members from Unitree, Zhiyuan, Xiaomi, Huawei, ZTE and Xpeng.* TechNode. **https://technode.com/2025/11/25/china-unveils-humanoid-robot-standards-committee-with-members-from-unitree-zhiyuan-xiaomi-huawei-zte-and-xpeng/**

[88] Beijing Municipal Government. (2025, April 24). *China's First National Standards for Humanoid Robots Approved for Development.* english.beijing.gov.cn. **https://english.beijing.gov.cn/beijinginfo/sci/event/202504/t20250424_4073087.html**

[89] *ANSI/A3 R15.06-2025 Robot Safety.* ANSI Blog. **https://blog.ansi.org/ansi/ansi-a3-r15-06-2025-robot-safety/**

[90] Udavant, S. (2025, November 24). *Cobots are out, collaborative applications are in after safety standard change.* Manufacturing

Dive. **https://www.manufacturingdive.com/news/cobots-iso-ansi-collaborative-applications-manufacturing/805970/**

[91] Bivans, N. (2025, November 11). *Humanoids - Safety Standards for the Next Wave of Robots.* RoboticsTomorrow. **https://www.roboticstomorrow.com/article/2025/10/humanoids-safety-standards-for-the-next-wave-of-robots-/25631**

[92] Novanta Inc. (2025, October 24). *Novanta Hosts ISO event for Humanoid Robot Safety Standards.* Novanta Investor Relations. **https://investors.novanta.com/news/news-details/2025/Novanta-Hosts-ISO-event-for-Humanoid-Robot-Safety-Standards/default.aspx**

[93] Leung, A. Y. M., Zhao, I. Y., Lin, S., & Lau, T. K.. (2022, December 22). *Exploring the Presence of Humanoid Social Robots at Home and Capturing Human-Robot Interactions with Older Adults: Experiences from Four Case Studies.* Healthcare (Basel). **https://pmc.ncbi.nlm.nih.gov/articles/PMC9818881/**

[94] Yole Group. (2025, November). *Humanoid robots 2025: The race to useful intelligence.* Edge AI + Vision. **https://www.edge-ai-vision.com/2025/11/humanoid-robots-2025-the-race-to-useful-intelligence/**

[95] Alexander, S. (2020, January 30). *Book Review: Human Compatible.* Slate Star Codex. **https://slatestarcodex.com/2020/01/30/book-review-human-compatible/**

www.ingramcontent.com/pod-product-compliance
Lightning Source LLC
Chambersburg PA
CBHW071737200326
41519CB00021BC/6762